The Epistemic Life of Groups

MIND ASSOCIATION OCCASIONAL SERIES

This series consists of carefully selected volumes of significant original papers on predefined themes, normally growing out of a conference supported by a Mind Association Major Conference Grant. The Association nominates an editor or editors for each collection, and may cooperate with other bodies in promoting conferences or other scholarly activities in connection with the preparation of particular volumes.

Director, Mind Association: Julian Dodd
Publications Officer: Sarah Sawyer

Recently Published in the Series:

Reality Making
Edited by Mark Jago

The Metaphysics of Relations
Edited by Anna Marmodoro and David Yates

Thomas Reid on Mind, Knowledge, and Value
Rebecca Copenhaver and Todd Buras

The Highest Good in Aristotle and Kant
Joachim Aufderheide and Ralf M. Bader

Foundations of Logical Consequence
Edited by Colin R. Caret and Ole T. Hjortland

The Highest Good in Aristotle and Kant
Edited by Joachim Aufderheide and Ralf M. Bader

How We Fight: Ethics in War
Edited by Helen Frowe and Gerald Lang

The Morality of Defensive War
Edited by Cécile Fabre and Seth Lazar

Metaphysics and Science
Edited by Stephen Mumford and Matthew Tugby

Thick Concepts
Edited by Simon Kirchin

The Epistemic Life of Groups

Essays in the Epistemology of Collectives

Edited by
Michael S. Brady and Miranda Fricker

OXFORD
UNIVERSITY PRESS

Great Clarendon Street, Oxford, OX2 6DP,
United Kingdom

Oxford University Press is a department of the University of Oxford.
It furthers the University's objective of excellence in research, scholarship,
and education by publishing worldwide. Oxford is a registered trade mark of
Oxford University Press in the UK and in certain other countries

© the several contributors 2016

The moral rights of the authors have been asserted

First Edition published in 2016

All rights reserved. No part of this publication may be reproduced, stored in
a retrieval system, or transmitted, in any form or by any means, without the
prior permission in writing of Oxford University Press, or as expressly permitted
by law, by licence or under terms agreed with the appropriate reprographics
rights organization. Enquiries concerning reproduction outside the scope of the
above should be sent to the Rights Department, Oxford University Press, at the
address above

You must not circulate this work in any other form
and you must impose this same condition on any acquirer

Published in the United States of America by Oxford University Press
198 Madison Avenue, New York, NY 10016, United States of America

British Library Cataloguing in Publication Data

Data available

Library of Congress Control Number: 2915946838

ISBN 978-0-19-875964-5

Links to third party websites are provided by Oxford in good faith and
for information only. Oxford disclaims any responsibility for the materials
contained in any third party website referenced in this work.

Contents

List of Contributors	vii
Introduction *Michael S. Brady* and *Miranda Fricker*	1

Part I. Epistemology

1. Mutuality and Assertion Sanford C. Goldberg	11

2. Fault and No-Fault Responsibility for Implicit Prejudice:
   A Space for Epistemic 'Agent-Regret'
   Miranda Fricker
	33

3. On Knowing What We're Doing Together: Groundless Group
   Self-Knowledge and Plural Self-Blindness
   Hans Bernhard Schmid
	51

Part II. Ethics

4. The Social Epistemology of Morality: Learning from the
   Forgotten History of the Abolition of Slavery
   Elizabeth Anderson
	75

5. Group Emotion and Group Understanding Michael S. Brady	95
6. Changing Our Mind Glen Pettigrove	111

Part III. Political Philosophy

7. The Epistemic Circumstances of Democracy Fabienne Peter	133

8. The Transfer of Duties: From Individuals to States
   and Back Again
   Stephanie Collins and Holly Lawford-Smith
	150

9. Four Types of Moral Wriggle Room: Uncovering Mechanisms
   of Racial Discrimination   173
   Kai Spiekermann

Part IV. Philosophy of Science

10. Collective Belief, Kuhn, and the String Theory Community   191
    James Owen Weatherall and *Margaret Gilbert*

11. Collaborative Research, Scientific Communities, and the
    Social Diffusion of Trustworthiness   218
    Torsten Wilholt

Bibliography   235
Index   251

List of Contributors

Elizabeth Anderson	University of Michigan, Ann Arbor
Michael S. Brady	University of Glasgow
Stephanie Collins	University of Manchester
Miranda Fricker	University of Sheffield
Margaret Gilbert	University of California, Irvine
Sanford C. Goldberg	Northwestern University
Holly Lawford-Smith	University of Sheffield
Fabienne Peter	University of Warwick
Glen Pettigrove	University of Auckland
Hans Bernhard Schmid	University of Vienna
Kai Spiekermann	London School of Economics
James Owen Weatherall	University of California, Irvine
Torsten Wilholt	Leibniz Universität Hannover

Introduction

Michael S. Brady and *Miranda Fricker*

Groups engage in epistemic activity all the time—whether it be the active collective inquiry of localized epistemic communities such as scientific research groups or crime-detection units, or the heavily institutionally structured evidential deliberations of tribunals and juries, or the more spontaneous and imperfect information-processing of the voting population. In the philosophy of mind and action there is a matured literature advancing competing theories of what groups are and how they do what they do.[1] Such debates principally turn on whether groups are best construed as no more than the sum of the individuals that compose them acting in concert, or whether certain forms of practical and intentional interdependencies suffice to generate a new agent, a distinct group agent that is greater than, or at least different from, the sum of its parts. More recently, social epistemology has also flourished, expanding and making connections with other areas of philosophy.[2] With these two philosophical lenses now beginning to align themselves, the new vista before us is that of *collective epistemology*—a natural next step for social epistemology.[3]

One of the key aspects of group agency is the deliberation that lies behind it, and the various epistemic commitments and capacities that are involved in such deliberations. The relevant debates in epistemology, concerning such things as competing accounts of collective belief, justification, and knowledge, have now begun to flourish. The

---

[1] A few indicative landmarks in this expansive territory might be Gilbert (1989 and 2000); Bratman (1999); List and Pettit (2011); Tuomela (2013).

[2] See e.g. recent edited collections in social epistemology: Goldman and Whitcomb (2010); Haddock, Millar, and Pritchard (2010). Also Grasswick and Webb (2002), which was part of an issue of *Social Epistemology* devoted to the theme of 'Feminist Epistemology as Social Epistemology'; in which general connection, see also Grasswick (2013).

[3] To track the rise of 'collective epistemology', see Gilbert (2004), and Tollefsen (2007a); the special issue of *Social Epistemology* edited by Mathieson (2007); and most recently collections edited by Schmid, Sirtes, and Webe (2011) and Lackey (2014).

essays in this book, however, are not on the whole directly focused on these issues, but rather explore different epistemic aspects of the behaviour of different sorts of group—institutional bodies, the moral community, informal groups, religious communities, the state, or simply the population at large. To this extent the book is part of an expansionist trend in epistemology of the last decade, consisting in the exploration of new epistemological projects that go beyond the traditional problems such as the refutation of scepticism, the nature of warrant, or the analysis of knowledge. The essays collected here explore different aspects of the epistemic practices of groups, thereby indicating the great range of ways in which epistemological issues permeate the well-functioning, or otherwise, of different kinds of human collectivity.

This volume collects essays by leading philosophers of different but coinciding persuasions in order to generate a more multilateral conversation than is ordinarily possible—but which is highly desirable given the manifestly common concerns, and the breadth of significance associated with these matters of collective epistemic practice. The essays each explore some region of our collective epistemic practice; and each essay has a particular focus that brings it under one of the following broad section headings: Epistemology, Ethics, Political Philosophy, and Philosophy of Science. The essays in the Epistemology section (Part I) address topics that make them fit squarely within the core social epistemological remit; the essays in the other sections address epistemological strands running through topics that primarily belong in other subject areas of philosophy. Together the essays indicate the richness and complexity of the philosophical issues generated by thinking about the epistemic life of groups. In what follows we give a brief outline of the individual chapters, before closing with remarks about some of the central issues raised.

Part I. Epistemology

Sanford C. Goldberg's chapter, 'Mutuality and Assertion', illustrates the importance of collective epistemology to a core topic in recent epistemology, namely epistemic disagreement. Goldberg's central question is this: can assertions be warranted under conditions of systematic disagreement—disagreement of the sort we find in philosophy, politics, religion, and the more theoretical parts of the social and natural sciences? On the one hand, he argues, there are strong reasons to regard assertion as governed by a demanding epistemic norm (such as knowledge), and it is plausible to think that in contexts of systematic peer disagreement we sometimes (often?) fail to attain knowledge. On the other, the practice of assertion persists in these areas, even under conditions of systematic peer disagreement. Indeed, the discipline of philosophy would appear to depend on this practice. Goldberg proposes that this tension can be resolved by appeal to the hypothesis that the standard set by assertion's norm is fixed in part in terms of something beyond the individual: namely, what is mutually believed by the speaker and her audience in the context in which the assertion is made. This is what he terms the *Mutual Belief Norm*. His chapter aims to provide independent grounds for this hypothesis. His central argument is that we can do so by appeal to Grice's guiding idea that conversation is a cooperative activity between agents, so that the rationality of speech contributions is to be understood by reference to the group context in which they are operating.

Miranda Fricker's chapter, 'Fault and No-Fault Responsibility for Implicit Prejudice: A Space for Epistemic "Agent-Regret"', explores different forms of epistemic responsibility, individual and collective, as regards the influence of prejudice on judgement. On the whole, if one makes judgements that are significantly influenced by prejudicial bias, then one is epistemically at fault, so that epistemic blame would be justified, including self-blame. What about cases where the prejudice in question is an 'implicit bias' (non-conscious, automatic, evidence-resistant, and possibly contrary to one's beliefs)? Here too, Fricker argues, the default is that we stand as blameworthy, though allowing that there may often be extenuating circumstances that diminish the degree of appropriate blame. Compare an entrenched character trait of which the subject is unaware partly because it does not fit with her self-conception—'implicit' selfishness, perhaps.

Fricker asks, however, whether there are circumstances in which we are guilty of implicit prejudice and yet where blame *is* entirely inappropriate (not merely extenuated). An example might be a case of *environmental epistemic bad luck*: where there is prejudice in the epistemic environment, *and* one has no reason to suspect that this is so, resulting in an *epistemically innocent* inheritance of environmental prejudice. Where this is so, argues Fricker, we confront the space of a 'no-fault epistemic responsibility'—the epistemic analogue of 'agent-regret'. We are not epistemically blameworthy, and yet we still have responsibility, as is revealed in the fact that there are epistemic obligations which apply to us specifically because it was through *our* epistemic agency that the prejudiced judgement was made. The fulfilment of those epistemic obligations, it is argued, will typically require the promotion of collective institutional measures to militate against the influence of prejudice in the future, This will be the case insofar as the practical remit of the individual's obligation coincides with an existing responsibility of the institution—for instance, where an individual employee has responsibility for a given promotions process with the organization. This coincidence of areas of responsibility means that the individual's responsibility to ameliorate a situation of potential bias quickly generates a collective, organizational one. And so, Fricker concludes, any counter-biasing epistemic responsibilities of individuals acting under the auspices of an institutional body (as an employee, for instance) will tend to beget collective epistemic responsibilities to take ameliorative counter-biasing measures.

Hans Bernhard Schmid contributes the third chapter to this section: 'On Knowing What We're Doing Together: Groundless Group Self-Knowledge and Plural Self-Blindness'. Schmid is concerned with whether an influential view about intentional behaviour at the individual level is also true at the collective level. This is the view that in order to act intentionally, an agent needs to know what she is doing, where this knowledge is 'groundless'—that is, non-observational and non-inferential. His central question is this: is our knowledge of what we are doing *together with others* of the same groundless kind? Schmid begins by highlighting the central features of groundless self-knowledge on Anscombean lines: these are first-person identity, first-person perspective, first-person commitment, and first-person authority. He then considers an argument that such knowledge is not available at the group level, on the grounds that a requirement for group knowledge is that each individual needs to know what her partners are doing, and this would seem to require observation and inference. Schmid concludes, however, that a moderate version of the claim that there is groundless group self-knowledge is defensible, if we maintain that the relevant knowledge involves plural pre-reflective and non-thematic self-awareness of what it is that the group members are jointly doing. Thus there is indeed a common structure to individuals' and groups' knowledge of their actions.

Part II. Ethics

Elizabeth Anderson's chapter, 'The Social Epistemology of Morality: Learning from the Forgotten History of the Abolition of Slavery', opens the section on Ethics. Anderson's concern is how social groups learn from history, and how the organization of social groups bears on the prospects for improvements to group beliefs. Anderson's focus is on the particular history of slavery. She notes that during the nineteenth century, the belief that individuals have a right against being enslaved became a nearly worldwide consensus. Most people today believe that this change in moral convictions was a case of moral learning. But Anderson is concerned with how we can know this, or similar claims about moral progress, without begging the question in favour of our current beliefs. She proposes to answer this question by developing a naturalized, pragmatist moral epistemology through case studies of moral lessons people have drawn from the history of abolition and emancipation. Anderson argues that processes of contention, in which participants challenge existing moral and legal principles governing interpersonal relationships, play critical roles in moral learning. Contention may take the form of argument, but it takes many other forms as well, including litigation, protest, and revolution. Anderson concludes that progress in moral inquiry requires that groups are receptive to and open to the perspectives of others, and not simply of those in authority; it requires 'the practice of epistemic justice by and for all'.

Collective understanding is the focus of Michael Brady's contribution to this volume, 'Group Emotion and Group Understanding'. Brady aims to explain how group emotion can have positive epistemic value in so far as it promotes group understanding; and he argues that this epistemic good would be very difficult to achieve, in many cases, without group emotion. After outlining philosophical, psychological, and neuroscientific support for the view that individual emotion promotes individual understanding, Brady applies this picture to group emotion and group understanding, and illustrates the connection between the two by focusing on the phenomenon of *public inquiries*. On this view, these inquiries are both motivated by group emotion, and aim at the attaining and dissemination of information so that some group understands what has happened. Without group emotion, he argues, it is unlikely that this collective epistemic good would be achieved. If this is correct, then group emotion is more valuable, from an epistemic standpoint, than traditional thinking on this issue supposes.

In 'Changing Our Mind', Glen Pettigrove considers the ways in which groups revise their beliefs, and proposes an account of group belief revision that allows modifications along a number of different dimensions. In particular, Pettigrove proposes an account of group belief revision that can accommodate modifications of (i) propositional content, (ii) non-propositional content, (iii) understanding, and (iv) conception. He develops his account by focusing on communities that are less discussed in the literature on social epistemology, namely moral and religious communities. By focusing on these communities, Pettigrove argues that Margaret Gilbert's account of group belief revision needs to be supplemented: while the view is adequate to changes in collective propositional knowledge-that, it fails to capture or accommodate revision of collective holistic knowledge-that, and in particular cannot accommodate revision to belief in groups such as religious communities or charitable organizations, which are built around normative commitments.

Part III. Political Philosophy

Fabienne Peter's chapter, 'The Epistemic Circumstances of Democracy', focuses on a long-standing question in political philosophy, namely: 'does political decision-making require experts, or can a democracy be trusted to make correct decisions?' Peter notes that the traditional debate about this issue is *instrumentalist*, in so far as it is thought that the determining factor for the legitimacy of political institutions is the epistemic status of the outcomes of decision-making processes. Supporters of democracy argue that this system produces optimal outcomes and hence can be trusted, whilst critics of democracy argue that outcomes of democratic decision-making are sub-optimal and hence the system cannot be trusted. Peter argues—against the instrumentalist approach—that attempts to defend democracy on epistemic instrumentalist lines are self-undermining. She proceeds to develop an alternative, procedural, epistemic defence of democracy, arguing that there is a *prima facie* epistemic case for democracy whenever there is no procedure-independent epistemic authority available on the issue that is to be decided.

In 'The Transfer of Duties: From Individuals to States and Back Again', Stephanie Collins and Holly Lawford-Smith consider whether a standard model for the transfer of duties from individuals to collectives applies to states' duties. On the standard model, individuals sometimes pass their duties on to collectives, which is one way in which collectives can come to have duties. This involves certain crucial epistemic transactions: notably, that the individual communicate to the collective the knowledge that she wills the transfer of duty; and that the individual makes all reasonable efforts to bring it about that she has a reasonable belief that the collective will indeed discharge the duty appropriately. The collective discharges its duties by acting through its members, which involves distributing duties back out to individuals. Individuals put duties in and get (transformed) duties out. But can this general account make sense of *states'* duties? And if so, to what extent? Do some of the duties we typically take states to have come from individuals having passed on certain individual duties? The authors note that there are complications to the picture: states can discharge their duties by contracting fulfilment out to non-members; states seem able to dissolve the duties of non-members; and some duties of states are not derived in this way. They argue that these complications do not undermine the general account and its application to states. Furthermore, Collins and Lawford-Smith show that the application has an interesting upshot: by asking which individuals *robustly* participate in this process of duty transfer-and-transformation with a given state, they show how we can begin to get a grip on who counts as a *member* of that state.

A different political problem is taken up by Kai Spiekermann in his chapter, 'Four Types of Moral Wriggle Room: Uncovering Mechanisms of Racial Discrimination'. Spiekermann describes recent experiments in behavioural economics which reveal that individuals frequently use so-called 'moral wriggle room' to avoid complying with costly normative demands. 'Wriggle room' describes our tendency to shape and modify our belief-system so as to convince ourselves that our behaviour is morally appropriate or satisfactory, and to thereby avoid stringent moral obligations. Spiekermann develops a typology of 'moral wriggle rooms' that helps to illustrate different opportunities for strategic information manipulation, and shows how moral wriggling can often operate in an unconscious, yet systematic way. He then notes that the experimental literature tends to be focused on individual behaviour; however, failures to meet obligations of inquiry are rooted in social practices and institutions. For example, one's individual ability to maintain biased beliefs is much higher when all of one's peers have the same biased beliefs. As a result, problems generated by moral wriggling must be addressed at the level of the group. Spiekermann illustrates this issue for social moral epistemology using the case study of racial discrimination, and considers a number of options we might employ to mitigate this problem.

Part IV. Philosophy of Science

James Owen Weatherall and Margaret Gilbert's chapter, 'Collective Belief, Kuhn, and the String-Theory Community', opens the final section of the volume, in which scientific aspects of collective epistemology are discussed. The chapter begins with Gilbert's well-known account of collective belief: this involves a joint commitment of certain parties, who constitute what she refers to as a plural subject. Gilbert has, in previous work, argued that ascriptions of beliefs to scientific communities commonly involve appeal to collective belief understood in her sense. This raises a potential problem when it comes to scientific change, however. For if Gilbert's view of collective belief as involving joint commitment is correct, and some of the belief ascriptions in question are true, then the members of some scientific communities have obligations that may act as barriers both to the fair evaluation of new ideas and to changes in scientific consensus. The authors argue that this may help to explain Thomas Kuhn's observations on 'normal science', and go on to develop the relationship between Gilbert's proposal and several features of a group of physicists working on a fundamental physical theory called 'string theory', as described by physicist Lee Smolin. Weatherall and Gilbert argue that the features of the string theory community that Smolin cites are well explained by the hypothesis that the community is a plural subject of belief. As a result, reflection on the practices of an actual scientific community provides further support for Gilbert's account of group belief.

The final chapter focuses on issues of trustworthiness in scientific research. In 'Collaborative Research, Scientific Communities, and the Social Diffusion of Trustworthiness', Torsten Wilholt argues for the thesis that when we trust the results of scientific research, that trust is inevitably directed, at least in part, at collective bodies rather than at single researchers. As a result, he proposes that reasonable assessments of epistemic trustworthiness in science must attend to these collective bodies. Wilholt supports his thesis by first explaining the collaborative nature of most of today's scientific research. He argues that the trustworthiness of a collaborative research group does not supervene on the trustworthiness of its individual members, and points out some specific problems for the assessment of epistemic trustworthiness that arise from the specific nature of today's collaborative research. Wilholt then argues that the social diffusion of trustworthiness goes even further; on his view, we always also need an assessment of the trustworthiness of the respective research community as a whole. Communities, he claims, play an essential role in the epistemic quality management of science. To see why this role is indispensible, Wilholt investigates and delineates three dimensions of what is desirable in a method of inquiry: the reliability of positive results, the reliability of negative results, and the method's power. Every methodological choice involves a trade-off between these three dimensions. The right balance between them depends on value judgements about the costs of false results and the benefits of correct ones. Conventional methodological standards of research communities impose constraints on these and thereby harmonize the implicit value judgements. Trusting that the research community has done this in a suitable way is thus always part of placing our trust in a scientific result.

Together, these studies indicate the extraordinary reach and internal diversity of the emerging field of Collective Epistemology. The strange individualism of English language philosophy is a historical contingency with a complex aetiology, and it has impacted on many different areas of philosophical discussion in ways that are not always easy to diagnose. But, in addition, it has also simply concealed many important philosophical questions from view, and one such is the range of questions addressed by the authors contributing to this book—questions relating to collective epistemic practice and its significance for how we know everyday things, how we pursue a shared moral life, how we conduct ourselves in professional contexts, how we design and modify our political institutions, and the forms of inquiry that govern the advancement of science. There are, of course, many other possibilities for collective epistemology than are exemplified here. Our hope is that this book will help further extend philosophy's conception of the proper domain of epistemology, thereby opening up many new promising avenues of understanding.

PART I

Epistemology

1
Mutuality and Assertion

Sanford C. Goldberg

1. The Problem: Assertoric Practice and Assessment

The aim of this chapter is twofold. First, I aim to develop a context-sensitive account of the norm of assertion in which the requirement set by this norm, taken to be epistemic in its content, crucially involves the mutual expectations of interlocutors within an epistemic community. Second, I aim to provide independent grounds for thinking that this account is true, and I aim to defend it against various objections.

Before doing either of these things, I want to present a problem, to which such an account (if independently motivated) would be a solution. The problem is that there are two background assumptions, each plausible in its own right (if not universally endorsed), yet which are incompatible with one another.

The first assumption concerns the norm of assertion itself. It is plausible to think that assertion has an epistemic norm E: speaker S's assertion that p is proper only if S satisfies E with respect to p, i.e. only if S has the relevant warranting authority regarding p. What is more, it is plausible to think that E is a demanding epistemic standard. The most prevalent view in the literature is that it is knowledge. If this is correct, then the norm of assertion tells us this: one should not assert that p, unless one knows that p.[1] But whether or not one thinks

---

I would like to thank Miranda Fricker and Michael Brady for very helpful comments on earlier versions of this chapter. I would also like to thank audiences at the various places at which I have given this paper: the philosophy departments at Monash University, the University of New South Wales, LOGOS (University of Barcelona), Vanderbilt University, the University of Cologne, and the University of Aarhus; and also the 2012 Orange Beach Epistemology Workshop on Social Epistemology, the 2011 Kentucky Philosophical Association meeting, and the 2011 Conference on 'Collective Epistemology: The Epistemic Life of Groups', at the Institute of Philosophy, University of London.

[1] This view has been endorsed by Unger (1975), Williamson (1996), DeRose (1996), Hawthorne (2003), and Stanley (2005).

that knowledge is the norm, it is widely agreed that the standard itself, E, is demanding.

The second assumption concerns the practice of assertion. In particular, there are plenty of cases in which, whether owing to pervasive disagreement (philosophy, politics, theoretical areas in the social and natural sciences, etc.) or low epistemic expectations (difficulties of achieving *knowledge* in highly theoretical areas), few if any speakers have any epistemic credentials such as knowledge, or knowledge-sufficient justification or warrant, to make assertions in these areas. And yet assertions continue to be made, unabated, in these areas. Philosophy is a particularly interesting domain in which to develop this problem. Elsewhere[2] I have argued that disagreement in philosophy is a special case of peer disagreement, where, whatever one thinks about the epistemic significance of disagreement as such, the sort of *systematic* disagreements one finds in philosophy make the prospects for epistemically high-grade belief remote.[3] Insofar as there is no epistemically high-grade belief at all under these conditions—a conclusion for which I argue in Goldberg (2013a)—I will call these conditions the conditions of *diminished epistemic hopes*. Yet assertions continue in philosophy, even under conditions of systematic disagreement, and hence diminished epistemic hopes. Indeed, the practice of philosophy would appear to depend on this.[4]

Taken together, these two assumptions suggest an unhappy conclusion: insofar as the practice of assertion continues even under conditions of diminished epistemic hopes, these assertions are *systematically improper*. In short, we appear to face a stark choice: accept the unhappy conclusion, or else reject one of the two assumptions. Yet none of these options seems particularly happy.

Consider first the option to accept the unhappy conclusion. We then face a situation in which every assertion made under conditions of diminished epistemic hopes is improper. If I am right that the practice of philosophy depends on the continued making of assertions under such conditions, then we would reach the unhappy conclusion that the practice of philosophy is at least to this extent suspect. Perhaps we might grant that, though assertions in question are improper, we let each other 'get away with' them anyway. But this suggestion is unhelpful: why would we think any higher of philosophy merely because we let each other get away with shoddy assertions? On the contrary, this would seem to condemn the practice twice-over: first, for improper assertions, second, for a

---

[2] Goldberg (2013a).
[3] By 'epistemically high-grade belief' I mean belief that amounts to knowledge or else is based on a knowledge-sufficient justification.
[4] See Goldberg (2013a and 2013b).

refusal to acknowledge this impropriety.[5] And what goes for philosophy goes for other areas in which the practice of assertion continues even under conditions of diminished epistemic hopes.

Consider next the option to reject the first assumption, to the effect that assertion has a demanding epistemic norm. To be sure, there are people who argue that assertion has no norm at all,[6] and others who argue that while assertion has a norm the norm is not epistemic.[7] Still, the vast majority of people who work on assertion seem to regard it as having an epistemic norm of some sort or other. Insofar as they are right, rejecting the hypothesis that assertion has an epistemic norm is already a cost. But now consider what must be done if, having accepted that assertion has an epistemic norm of some sort or other, we still want to reject the first assumption above. The conclusion would have to be that the norm is not particularly demanding. We would then have to replace the assumption of a *demanding* epistemic norm with an alternative account—one whereby the epistemic standard, E, is no more demanding than is needed in order to represent assertoric practice in these areas as desired. To be sure, there *are* unwarranted assertions under conditions of diminished epistemic hopes; the point above is rather that *not all* assertions made under these conditions are improper (i.e. some are proper). But even to get this result we would need to weaken E quite a bit. For if I am right that (in contexts of systematic peer disagreement) we face a situation of diminished epistemic hopes, then E would have to be weakened to the point where it demands something weaker than justified (or rational) belief. Of course, no sooner do we do this, than we have an obviously overly-permissive norm—one that sanctions a good deal of assertions we would want to regard as improper.

It is perhaps worthwhile discussing the sort of overly-permissive norm that emerges from our dialectic so far, if only to see why norms in this class would appear to be non-starters. To this end, consider the sort of norm that might be satisfied in conditions of diminished epistemic hopes. One candidate norm is that of mere belief. If the norm of assertion is mere belief, then a speaker might well satisfy it under conditions of epistemically diminished hope. But there would appear to be something curious in the idea that an assertion is warranted (proper, permissible) so long as the speaker *believes* what she says. For in effect this collapses (something like) the sincerity condition on assertion with assertion's

---

[5] Although the assertoric practice of philosophy is illustrative, it is not unique in this respect. Consider the theoretical parts of the social and natural sciences.
[6] See Pagin (2011). See also MacFarlane (2011) for an inventory of the sorts of views one can take about the nature of assertion.
[7] See Weiner (2005).

norm, and such a collapse would appear to have unacceptable implications. For example, it would follow that the incompetent believer is nevertheless a proper asserter (so long as he asserts only what he believes). One might think to rectify this by saying that it is a deficiency in assertion if one expresses what in fact is a deficient (unwarranted, unjustified) belief; such a view is presented by Bach (2008). But we would want to know why this is so: if an assertion is proper so long as the speaker believes the asserted content, it is unclear how the epistemic deficiency of the belief itself should affect the propriety of the assertion. After all, not all of the ways in which we would criticize an assertion bear on its propriety *qua assertion*: assertions that are rude, irrelevant, or made in too loud a voice are thereby criticizable, but if the assertion in question expressed the speaker's knowledge, then, at least with respect to the standards of the kind of speech act it was, the assertion was proper. Thus the proponent of the belief norm would appear precluded from saying that the assertion of an unjustified belief is improper; such a proponent is restricted to saying that the assertion is criticizable on *other* grounds, in violation of a standard other than that pertaining to the kind of speech act it was. This appears to me to be a great cost. Of course, what goes for the norm of belief also goes, *mutatis mutandis*, for other candidate norms that can be satisfied under conditions of epistemically diminished hopes. Such norms are too permissive.

But if we reject the option to weaken the norm of assertion to accommodate the cases in question, and if we also reject the option of accepting the unhappy conclusion that there are no proper assertions under conditions of epistemically diminished hope, we are left with the option to reject the second assumption above. This was the assumption that there are many areas in which the making of assertions continues even under conditions of systematic disagreement, and so even under conditions in which there is a recognition that the hope for knowledge and justified belief is remote.[8] To reject this assumption is to hold that assertions are not (typically, standardly) made in these circumstances after all. Why might one think this? Well, one might argue that what appear to be assertoric speech acts *aren't really assertions after all*. Perhaps they are hedged or qualified assertions; or perhaps they are a different speech act altogether. The difficulty with this reaction is that it seems seriously revisionary: although there surely *are* hedged or qualified assertions made in these contexts, it is also the case that (to all outward appearances) there are many examples of *straight assertions* in these contexts as well. To reject the second assumption is to have to treat these

---

[8] An initial reaction to this assumption might be to accept it, but hold that all such assertions are improper. This, in effect, would be to accept what I called the unhappy conclusion.

as cases in which things are not as they seem. I do not claim that this is unacceptable; only that, since it is revisionary, it would be better to see if we can address the problem without having to go this route.

With this in mind, I propose the following *desiderata* for a solution to our problem. We would like a solution to satisfy the following three conditions:

(1) it should not weaken E—the epistemic standard provided by the norm of assertion—to the point of irrelevance or insignificance;
(2) it should recognize the strengths of the arguments made on behalf of thinking that E is *knowledge*, or some other substantial epistemic property (such as knowledge-sufficient justification); and
(3) it should enable us to regard the relevant class of assertions as broadly proper despite the fact that few if any speakers have any substantial epistemic credentials to make assertions in these areas.

Now, it might be thought that there is no solution that satisfies all of (1)–(3). If this thought were true, then we would need to revisit the other options (dismissed above). However, in what follows I will argue that there is a solution that satisfies all of (1)–(3); and I will argue as well that this solution enjoys support independent of its offering a solution to the present problem.

2. Toward a Solution

The problem we are considering pertains to any domain in which the making of assertions continues despite diminished epistemic hopes. Above I mentioned one such domain—that of philosophy, where it is an ordinary part of the practice to make assertions under these conditions. I noted that it is tempting to describe this part of philosophical practice by saying that participants in conversations regularly let each other 'get away with' the assertions in question. Again, if this is merely a matter of letting one another 'get away with' what in fact are improper assertions, this is doubly bad for the practice. But we might well wonder whether this is the most perspicuous description of the practice. Perhaps it is not a matter of letting each other 'get away with' anything; perhaps the assertions in question are proper after all.

Of course, anyone who would like to try to make out such a view, and so who aims to satisfy *desideratum* (3) above, must face an immediate question: how to square the hypothesis that these assertions are proper after all, with the claim that few if any speakers in these areas enjoy any substantial epistemic standing with respect to the propositions they are asserting? In short, it can seem that the move to satisfy (3) will come at the cost of having to weaken the norm's standard to the point of irrelevance, and hence at the cost of failing to satisfy *desideratum* (2) above. I should emphasize as well that this difficulty has nothing in particular to do with assertoric practice in philosophy: it is a problem facing *any* discourse in which assertions continue to be made (as an ordinary part of the discourse) under conditions of diminished epistemic hopes.

In this section I argue that the difficulty can be met. I propose to do so by developing an account of assertion on which the standard provided by assertion's norm is set in a context-sensitive way. In this section I motivate such an account by appeal to broadly Gricean considerations; in the section following I argue that the account delivers the desired outcome, satisfying *desiderata* (1)–(3) above.

An account of assertion is an account of a type of speech act, and as such it should be presented against the background of our best understanding of speech acts. Paul Grice (1968/89) has provided an important grounding principle for such an understanding. Regarding speech as a rational, cooperative activity, Grice formulated the familiar Cooperative Principle as capturing a core part of the rationality of particular acts of this sort:

Cooperative Principle (CP)
Make your contribution such as it is required, at the stage at which it occurs, by the accepted purpose or direction of the talk exchange in which you are engaged.
(Grice 1968/89, 26)

Now most people who have employed Grice's CP (or the other elements of Grice's picture) have done so with an eye on characterizing the *content* dimension of communication. That is, they use Grice's framework to provide an account of how speakers manage to communicate more than they (strictly and literally) say, and of how hearers manage to recover what is communicated when this goes beyond what is (strictly and literally) said. But I see no reason why we can't use Grice's insight to shed light on the dimension of (illocutionary) *force*. This is what I propose to do.

One might wonder how Grice's CP can shed any light on the dimension of illocutionary force in general, and on assertion in particular. But this suggestion is not as outlandish as one might suppose. On the contrary, it is a natural one. To see this, consider that Grice went on to present various maxims that he regarded as falling out of CP. One of these was the maxim he labeled 'Quality', which I repeat here:

Quality (Q)
Do not say what you believe to be false. Do not say that for which you lack adequate evidence.   (Grice 1968/89, 27)

Now it is true that Grice's maxims are explicitly aimed at characterizing the notion of *saying* something. Still, when it comes to Q itself, it is not a far stretch to regard the maxim as contributing to our understanding of the more specific speech act of assertion. Indeed, we might well think that the 'quality' dimension of Grice's notion of saying just is a proposed characterization of the norm of assertion. On this picture, the speech act of assertion is governed by two rules: you shouldn't assert what you believe to be false, and you shouldn't assert that for which you lack adequate evidence. But precisely when is one's evidence 'adequate'? I submit that we should answer this question by appeal to the CP itself: the standards for adequacy of evidence are determined, at least in part, by 'the accepted purpose or direction of the talk exchange in which you are engaged'. If this is correct, then we have characterized a feature of the illocutionary force of an assertion by appeal to CP.

Still, we need an account of how 'the accepted purpose or direction of the talk exchange in which you are engaged' might serve to fix the standards of evidential adequacy. I propose that we can develop such an account in terms of Bach and Harnisch's useful (1979) notion of *mutual belief*. The following is their gloss on the role that mutual belief plays in the sort of inferences that are made in the course of the production and comprehension of speech:

Mutual Belief
If p is mutually believed between S and H, then (1) not only do S and H believe p, but (2) each believes that the other takes it into account in his thinking, and (3) each, supposing the other to take p into account, supposes the other to take him to take it into account.
(Bach and Harnisch 1979, 6)

I submit that the task of determining adequacy of evidence (and hence of determining the standards imposed by the maxim of Quality) is itself a special case of the sort of phenomenon of which Bach and Harnisch are speaking. In particular, if there is mutual belief that the hearer faces a practical task in which she is in need of information, and that she is relying on the speaker to provide this information, then adequate evidence would be the sort of evidence for a proposition which would render it reasonable for the hearer to act on the assumption that the proposition is true. If there is mutual belief to the effect that the hearer needs information of which she can be certain, then adequate evidence would be the sort of evidence that would support certainty.

The point I am presently making can be formulated in terms of the norm of assertion. Above I presented a schematic version of the hypothesis that assertion has an epistemic norm, in the form of the claim that:

ENA: S must: assert p, only if S satisfies epistemic condition E with respect to p, i.e. only if S has the *relevant warranting authority* regarding p.

My present claim employs the notion of mutual belief to address the matter of what epistemic standards one must satisfy if one is to count as having the relevant warranting authority regarding p. To a first approximation (to be modified in section 4 below), the hypothesis is this:

MBN: When it comes to a particular assertion that p, the relevant warranting authority regarding p depends in part on what is mutually believed by speaker and audience (regarding such things as the participants' interests and informational needs, and the prospects for high-quality information in the domain in question).

Below I will be arguing that (a slightly modified version of) this mutual belief model can address the challenge with which we began this chapter. My only claim here is that the model itself enjoys some independent support: it can be seen as deriving from a broadly Grice-inspired approach to speech exchanges.

This last point is worth dwelling on in a bit more detail. The model I am offering here is a special case of a more general picture of speech exchanges, one having nothing in particular to do with philosophy (or with disagreement, for that matter). On this picture, speech is a cooperative activity, and assertion is to be understood in these terms, as governed by rules of the sort Grice articulated in his principle of Quality. Insofar as these rules are themselves an object of (perhaps merely implicit) mutual belief, they determine a set of mutual expectations of speaker and hearer. That these expectations are (in part) epistemic, demanding adequacy of evidence, is precisely what makes assertion apt for playing the very important role it does: that of serving as the vehicle for the transmission of information. We can bring this out as follows. A hearer who observes a speaker make an assertion, under conditions in which the rules governing assertion are objects of mutual belief, will expect that the speaker acknowledges these rules, and so will expect the speaker to acknowledge the responsibility for having had adequate evidence. Insofar as the hearer regards the speaker as having succeeded at following the rules, then, the hearer regards the speaker as having adequate evidence; and when the hearer's so regarding the speaker is rational, this rationalizes the hearer's move to accept the information the speaker presented in her assertion, on the basis of her having so asserted.[9] In sum, it is because of the rules governing assertion that this speech act is apt for rationalizing hearers' beliefs in what is asserted—and precisely this renders assertion apt for the transmission of information.

---

[9] Of course, if the hearer was irrational in regarding the speaker as having conformed to the rules—the speaker asserted something regarding which it is common knowledge that no one has any evidence, or she had obvious vested interests in getting the hearer to believe what she said, etc.—then the hearer's acceptance is itself rationally flawed.

All of this is a familiar part of a broadly Grice-inspired picture of assertion. I am only introducing two wrinkles to this story. First, I am proposing that the determination of when evidence is 'adequate' to render an assertion proper is a matter of 'the accepted purpose or direction of the talk exchange'. Second, I am proposing to model the process by which this determination is made in terms of the notion of mutual belief. Both of these wrinkles (I have contended) are Gricean in spirit: they can be motivated as spelling out the picture of conversation Grice himself first introduced. I now want to put this picture to work in addressing the problem with which I began this chapter. In what follows my claim will be that what is mutually believed on a speech occasion sometimes includes information regarding not only (some subset of) the interests and informational needs of one's audience, but also the difficulties of acquiring epistemically high-quality information in a given domain in which the parties seek information. Insofar as this is the case, this body of mutual belief will affect the mutual expectations that speaker and hearer have of one another in the speech exchange. And it is this, I will go on to argue, that enables us to use the model to answer our original difficulty regarding philosophical assertion while simultaneously satisfying the three *desiderata* from the previous section.[10]

3. A Role for Epistemic Groups

Let us suppose that my two 'wrinkles' are correct: what counts as adequate evidence (sufficient to render assertion proper) is a matter of 'the accepted purpose or direction of the talk exchange'; and the purpose or direction of the talk exchange determined this through what is mutually believed in context. These two claims, by themselves, are not sufficient to address the problem with which I began this chapter—to account for the practice of assertion in the face of diminished epistemic hopes. The key point is easy to appreciate in connection with assertoric practice within philosophy. Insofar as it is mutual belief that sets the standard of the norm of assertion in a given context, then what standard is in play in a given context will depend on what is mutually believed in that context. But so far we have no guarantee that what is mutually believed in any arbitrary philosophy context will be such as to establish standards that are satisfied in those cases in which, intuitively, we want to say that the philosophical assertions are

---

[10] I should make it clear that while my proposed model has happy implications for the problem of philosophical assertion—the problem with which I began this chapter—it is proposed as modeling assertion as such.

proper. In short, unless there is a way to ensure that there is the relevant sort of mutual belief among philosophers, the standards that prevail in a given speech exchange between philosophers will be hostage to doxastic idiosyncracies of the speech participants. As noted, the problem here has nothing in particular to do with philosophy: the same sort of problem can be raised in any situation in which assertoric practice continues in the face of diminished epistemic hopes. If that practice is to be (broadly) proper, we will need a way to ensure that in each of the cases what is mutually believed in context sets the appropriate standard. How to ensure this? I believe that the answer lies in the notion of an *epistemic group*. After introducing this notion I go on to show that if a collection of people constitute an epistemic group, then this can be expected to have an effect on assertoric practice among them—and the resulting model of assertion will be able to satisfy *desiderata* (1)–(3) above.

I begin, then, with the notion of an epistemic group. It is sometimes mutual belief among participants to a conversation that they are speaking to one another *as neighbors*, or *as birdwatchers*, or *as philosophers*, or *as scientists*, or *as politicians*, or *as colleagues*, or... Being in such a group is sometimes associated with having certain kinds of knowledge (or at least belief). Such belief might regard the relevant subject-matter, the likely sources of further subject-matter information, the practices whereby such information is extracted, the ease of getting high-quality subject-matter information, and so forth. What is more, the existence of such knowledge or belief among the group's members can itself become a matter of mutual belief: the members of the group believe of themselves that each has the relevant subject-matter and procedural knowledge (or beliefs). When this is so, the members of such a group will use this mass of mutual belief to form mutual expectations: they will form expectations regarding what they can expect from each other, and what their group-interlocutors are likely to expect from them in turn, as they engage in communication and cooperative inquiry. When a group has such mutual (epistemic) expectations regarding each other as epistemic (knowledge-seeking) agents, I will call the group an *epistemic group*.

It is in the context of epistemic groups that we can reconsider the challenge to ensure that relevant mutual belief will prevail in any case involving philosophical assertion. In general, epistemic groups are the repository of mutual (subject-matter and inquiry-related) beliefs—and this background of mutual belief shapes assertoric practice within the group. The case of philosophy is a useful example.

Suppose that, in the course of making an argument for some controversial conclusion regarding (say) the nature of knowledge, Smith advances various claims—she asserts various propositions—as premises in her argument. She herself is well aware that with respect to each of the claims (assertions) she is making here, there will be those in the audience who will disagree with her. She could try to address all of their concerns, of course. But she knows, and anticipates that they know, that doing so would quickly spiral out of control: to vindicate each claim (assertion) in her main argument as itself the conclusion of a further argument, she would need to make still further claims (assertions), and she anticipates that for at least some of these (and perhaps many or all of them) there would be further disagreement. Since she takes all of this as mutually known by herself and her audience—in any case, since she takes it as reasonable to suppose that any well-trained philosopher will be aware of all of this, and aware that the others are aware of it too—she anticipates that her audience will appreciate the position she is in. She recognizes, of course, that one can't simply assert anything, willy-nilly, in philosophical contexts. But in light of what she reasonably regards as mutually believed among philosophers working in epistemology, she anticipates that her audience will expect that for any assertion she makes, she will be in a position to provide a defense of the assertion in question satisfying the standards for epistemology. Which is to say that the argument need not be knockdown, nor need it even persuade the other side (they can remain rational even as they deny her assertion); it need merely conform to professional standards (of rigor, clarity, responsiveness to reasons, etc.). This, then, is what informs her assertoric behavior.

What sort of evidence do we expect of one another when engaging in a speech exchange on some philosophical topic? It is not clear that there is one set standard for all of philosophy. On the contrary, it seems plausible to suppose that the expected standard can vary according to subject-matter: what we expect from a speaker who is advancing what she presents to be a theorem in logic is one thing, what we expect from a speaker who is advancing a claim in ethics (for example) is another still. I submit that this is because of what it is reasonable to assume is mutually believed among academically trained philosophers. It is reasonable to assume mutual belief among philosophers to the effect that propositions in logic can be established or refuted by proof. This is why we will expect one advancing such a claim (by way of making an assertion) to have evidence that amounts to, approximates, or stands in for a proof. It is also reasonable to assume mutual belief among philosophers to the effect that propositions in ethics cannot (typically) be established in this way. This is why we will not expect anything approximating a proof of someone advancing a claim in ethics. Rather, what we will expect in the way of evidence in ethics turns on what is mutually believed—better, what it is *reasonable to assume* is mutually believed[11]—regarding the nature of the subject-matter in ethics: the sorts of considerations that can support such claims, the difficulty of synthesizing all of the considerations bearing on a given ethical question, the nature of the sorts of methods we use in doing so, and so forth. If it is mutually acknowledged that a certain claim in ethics is part of a systematic disagreement, participants in the speech exchange will—or at any rate, should—adjust their evidence-related expectations accordingly. In particular, they should not expect the evidence to be decisive and clear; rather, they should expect that the evidence suffices to put the speaker in a position where she can 'make a case for' the claim in question. (Precisely what this amounts to is difficult to put into words, though I assume that philosophers will recognize the phenomenon.) And what goes for contested ethical propositions, goes more generally for contested philosophical propositions—at least insofar as the participants to the speech exchange have the sort of mutual belief I have described.

Of course, examples like that of Smith above, in which what is mutually believed among participants in a conversation helps to shape assertoric practice, can be found outside of philosophy as well. For one thing, the case of philosophy is a special case of the phenomenon of epistemic groups; there are other epistemic groups, and so we should expect other examples in which mutually recognized membership in a relevant epistemic group will delimit the assertion-generated expectations that speakers and hearers have towards one another. For another thing, cases of discussions between members of an epistemic group are themselves special cases in their own right: many, and perhaps most, of our conversations are not as members of an epistemic group. In those cases, mutual belief shared among participants will either be *ad hoc* (formed in context) or formed in some other way; but if my hypothesis is correct, these will still be cases in which what is mutually believed in context sets the standard of the norm of assertion.

In fact, it is with these sorts of case in mind, where there is no relevant epistemic group in play, that we can reconsider *desideratum* (2) above. Earlier I claimed that a norm of assertion built around MBN can satisfy this *desideratum*, and so can accommodate the strengths of the arguments made on behalf of thinking that the norm's standard is *knowledge* (or some other substantial epistemic property). I am now in a position to defend this claim, as follows. Having endorsed the Gricean claim that one should not assert that for which one lacks adequate evidence, my suggestion has been that adequacy of evidence ought to be determined in conjunction with the 'purpose of the talk exchange' itself.

---

[11] I will return to this in section 4, where I will use the 'reasonable to assume' condition to modify the MBN principle itself.

One way to understand this is that speakers ought to make their contributions ones that are helpful to their audience. Insofar as the audience's needs and interests are matters of mutual belief between speaker and hearer, these needs and interests help to determine both what information would be helpful to the hearer, and what sort of evidence would be needed to warrant an assertion of the relevant content. This is the case in the example of Smith above, in which two people speak to one another as members of a given epistemic group (academically trained philosophers). But now consider cases in which mutual belief is minimal: there is no relevant epistemic group in play, and no other basis for substantial mutual belief among participants in a speech exchange. In such cases, the speaker will not be in a position to tell either what information would be useful to the hearer, or what the hearer wants to do with it. In such circumstances, I submit, there will be upward pressure on the norm of assertion. This is for the simple reason that, in these circumstances, the speaker will have to make assertions which are such that, *no matter the hearer's informational needs*, the contribution is helpful to her. But this means that the speech contribution will have to be helpful even if the hearer's needs require a high epistemic standard to be met. Under these conditions, knowledge would appear to be required. We might model this point by saying that knowledge is the norm's 'default setting'—its setting when what is mutually believed is not robust enough to adjust the standard in any particular way. But whether we choose to model the point in this way, my present claim is simply that the context-sensitive account I am offering has the resources to accept that (and to explain why) knowledge is the relevant standard for assertion in a good number of the conversations in which we make assertions. The account thus can claim to capture much of what motivates the knowledge norm, and can indeed accept knowledge as the relevant standard in many, perhaps even most, contexts.

Indeed, it is possible to argue for the present proposal (to the effect that knowledge is the norm's 'default setting') from the reverse direction.[12] I have just suggested that it is the interlocutors' respective lack of knowledge of each others' interests and needs that exerts upward pressure on the standard, with the result that knowledge is the standard required. But we might start by noting that the typical case of assertoric speech exchange is one in which what interlocutors want from each others' assertions is knowledge. If this much is correct, then knowledge would be required unless particular expectations were somehow to affect this default. The suggestion here is that knowledge is the default standard

---

[12] With thanks to Miranda Fricker for the suggestion that follows.

from the outset, and is adjusted lower (or higher) only if there are overriding expectations in the context.

I confess that I am uncertain as to which is the better explanation. But this need not deter us here, since, in either case, we have an account on which knowledge is the 'default setting' of the norm of assertion. And this is all that we need to see how MBN itself, together with the hypothesis that philosophers constitute an epistemic group, can be used to respond to our original problem in a way that satisfies *desiderata* (1) and (3) as well.

Take *desideratum* (1) first, which is the desire not to weaken the epistemic norm of assertion to the point of practical irrelevance. I just argued that, by the lights of MBN, we would predict that in contexts of minimal or no mutual belief the prevailing standard will be as demanding as knowledge. In fact, MBN would predict that this standard will prevail in *any* context in which what is mutually believed is minimal.[13] In addition, in any case in which it is mutually acknowledged that the hearer is in need of knowledge, this will be the prevailing standard. To the extent that these cases are prevalent in ordinary, everyday life, MBN will not weaken the epistemic norm of assertion to the point of irrelevance—to the contrary, it will hold it up high.

Still, it might be wondered whether cases of minimal relevant mutual belief are prevalent in ordinary, everyday life. And here one might think that my position faces a dilemma. Either such cases of minimal mutual belief are prevalent, in which case my account can satisfy *desideratum* (1) (and so will keep the epistemic standard on assertion, E, substantial) but will fail to satisfy *desideratum* (3) (and so will fail to sanction the assertions that, intuitively, are warranted); or else such cases are not prevalent, in which case my account can satisfy *desideratum* (3) but not *desideratum* (1). In response, I want to highlight the context-sensitivity of MBN, together with the idea that assertoric practice within philosophy really is far from the standard instance of the production and consumption of assertion. The most common use for assertions is in satisfying others' informational needs; and the most common sort of need is a need (not just for information backed by good arguments and reasonable defenses, but rather) for *knowledge*. In this respect, what philosophers expect of one who advances a claim—namely, dialectically substantial reasons that are sensitive to the force of prevailing objections—is far from the standard expectations we have of assertions generally.

---

[13] Here I am silent as between the two explanations offered above for the phenomenon whereby knowledge is the norm whenever mutual belief is minimal—the one being that minimal mutuality exerts upward pressure on the standard, the other being that minimal mutuality is not a sufficiently robust basis for adjusting an already-in-place default of knowledge.

In most cases it suffices to establish the propriety of one's assertion that one has the corresponding piece of knowledge, whether or not one could provide substantial reasons on behalf of what was asserted, and whether or not one anticipated a host of objections. Assertions in philosophical contexts are, in this respect at least, far from ordinary.[14] But insofar as academically trained philosophers are an epistemic group, they will have a rich stock of mutual belief as setting their mutual expectations, and in this sort of context the standards do adjust. In this way MBN satisfies (3); and since it takes a good deal of mutual belief among participants to adjust the standards in this way, and since in other cases the standards remain high (unless adjusted by still further mutual belief), MBN satisfies (1) as well.

I conclude, then, that MBN offers the prospects for replying to our original difficulty, in a way that satisfies all of the *desiderata* (1)–(3). Unfortunately, it is not MBN as it stands, but a slightly modified version, that is defensible. I turn to this now.

4. Objections Considered

I have been arguing for MBN, according to which the standard for warranted assertion is fixed in a context-sensitive fashion, in terms of what is mutually believed in context. I have been trying to motivate this claim by appeal to a Gricean orientation on speech exchanges: the suggestion is that MBN can be seen as a special case of the Cooperative Principle, applied to the 'adequate evidence' condition in the sub-maxim of Quality. It is important to me that the proposal can be motivated in this way: it provides support that is independent of the support provided by the fact (if it is a fact) that MBN enables us to respond to our original difficulty while simultaneously satisfying *desiderata* (1)–(3). Still, I recognize that, as a thesis about the norm of assertion, MBN will not be uncontroversial. In this section I aim to address several objections that might be leveled against it. I will argue that, while several of them do point out some costs of the proposal, none of them succeeds in undermining it. Or rather, none of them succeeds in undermining a slight variant on MBN, the motivation for which will be apparent as we deal with the objections.

4.1. I begin first with an objection based on the contention that parties to a discussion cannot always tell whether there is a background of mutual belief. By itself, this contention need not amount to an objection: insofar as participants to

---

[14] With thanks to Michael Brady for indicating the need for this comment.

a speech exchange are uncertain whether there is mutual belief, they will not be warranted in proceeding on the assumption that there is mutual belief, in which case the 'default setting' for the norm will prevail. Still, one might try to work this contention into an objection by considering cases in which there is (not merely uncertainty but) disagreement in the respective parties' beliefs regarding what is mutually believed. Cases of this sort might be thought to raise an objection to MBN, on the grounds that MBN can offer no plausible verdict on such cases.

Actually, I think that, far from constituting an objection to MBN, such cases can actually be used to support MBN: not only does MBN offer verdicts in such cases, what is more, the verdicts it yields are independently plausible. Take a situation in which one or more parties to a discussion has/have false beliefs regarding what is mutually believed in context. In such cases the two sides will not be calibrated in their respective expectations regarding the assertions being made by the other side. It will be helpful to have a schematic case before us. Suppose that speaker S, taking herself to be speaking to fellow philosophers on a topic which (she thinks) all will acknowledge as deeply controversial, holds that the standards in play will be those governing philosophical exchanges. Accordingly, S's assertions are governed by a 'reasonable by philosophical standards' norm; and accordingly, S expects that H's expectations regarding S's assertions will be adjusted accordingly. However, audience H's expectations have not been calibrated to S's. On the contrary, H regards the context as one in which there is little or no relevant mutual belief. Consequently, H expects that, given the minimal mutual belief, S's assertions must meet suitably high epistemic standards (knowledge perhaps). (Assume that H cannot tell, merely from S's assertion, that the context is one of philosophy.) Clearly, S's and H's mutual expectations are not aligned; this is a case involving a failure of calibration.

Now let us ask: supposing S to assert that p under conditions of this sort of calibration failure, is S's assertion warranted or not? The objection we are presently considering holds that the proponent of MBN has nothing plausible to say in answering this question. But this is not so. I submit that, in cases of calibration failure, the question 'Warranted or not?' cannot be answered prior to determining the justificatory (epistemic) status of the beliefs that generated S's and H's respective expectations—and this will depend, in turn, on the prevalence and salience of the relevant groups and group practices. For example, suppose that S and H are at a philosophy convention, and that S's assertions are of propositions which anyone educated in philosophy would be aware are philosophical in nature; then it would seem that S's underwriting beliefs—the beliefs underwriting S's expectations regarding H—are reasonable, whereas H's underwriting beliefs—the beliefs underwriting H's expectations regarding S—are not.

In that case, we can say this: while S's beliefs regarding what was mutually believed were false, still, they were reasonable; whereas H's beliefs regarding what was mutually believed were not merely false but unreasonable; and so, regarding the standard to apply to the case, we favor what S took to be mutually believed. To be sure, endorsing this verdict means replacing MBN with a view according to which what it is *reasonable* to regard as mutually believe is what sets the standard:

RMBN: When it comes to a particular assertion that p, the relevant warranting authority regarding p depends in part on *what it would be reasonable for all parties to believe* is mutually believed among them (regarding such things as the participants' interests and informational needs, and the prospects for high-quality information in the domain in question).

Given RMBN's standard, S's assertion was proper (supposing, of course, that it met the threshold of philosophical defensibility). Of course, if the situation were otherwise—S and H don't know each other, they meet up on the street, there are no grounds to assume that they share a philosophical background, etc.—then it would be H's underwriting beliefs that are reasonable, not S's. In this case we should conclude that the standard of assertion's norm is fixed by what H took to be mutually believed, and so S's assertion was unwarranted.

It is worth noting that RMBN can be motivated in Gricean terms in precisely the way that MBN itself was. Insofar as speech is a rational, cooperative activity, it stands to reason that the standards ought to be those that reasonable practitioners follow. In saying this, we abstract away from what speech participants actually believe regarding what is mutually believed in context, to what they *ought* (reasonably) to believe is mutually believed in context. One might wonder how to determine what ought to be believed in this respect. This is a fair question. But this should be no more concerning to us than is the question when 'evidence' counts as 'adequate'—a question that any Gricean will have to face. In this manner we can see that, while the move from MBN to RMBN does involve a modification, it is no departure from the Gricean picture itself.[15]

Still, RMBN itself begs a question: what should be said of cases in which both side's underwriting beliefs are (not merely false) but equally reasonable (alternatively: equally unreasonable)? For surely there can be such cases. Simply imagine that, although they are in the lobby of a hotel that is currently hosting a philosophy conference, still there are many people other than philosophers mingling there; neither S nor H dress in any way that identifies them as philosopher or

---

[15] I am indebted to Miranda Fricker for the suggestion that I make the present point.

non-philosopher; H is not aware that there is a philosophy conference going on, and this ignorance is not unreasonable; and so forth. Then if S, taking H to be a philosopher, starts talking philosophy, it seems that their case is one in which both sides' underwriting beliefs are both false and equally reasonable. Here I think that the proponent of RMBN should simply acknowledge that there may be no fact of the matter regarding the propriety of the assertion. Nor do I regard this as an unhappy thing to have to say; I think that it captures a real fact, namely, that under conditions in which two sides err regarding what they mutually believe, where both sides' underwriting beliefs are equally reasonable, it seems simply wrong to think that there is some fact of the matter that determines the relevant standard to apply to the speaker's assertion.

4.2. Still, a related objection remains. This objection—better, this bundle of objections—derives from the fact that in many conversations there are multiple epistemic groups with which individual participants might identify in a given case. The point can be made about speakers, audiences, or both. Thus a speaker might be addressing her audience as a politician, as a citizen of this town or this region, etc., and/or as a member of a particular profession; and there might be different possible candidate mutual beliefs corresponding to each of these groups. What is more, her audience might identify themselves in any of these or other ways, again with different possible candidate mutual beliefs corresponding to each. But there is still more: audiences often include many people, where there is no single epistemic group with which all identify, and where most in the audience identify with many different epistemic groups. In such complicated cases, what determines the mutual belief that sets the norm of assertion? Again, this turns from a question into an objection if the proponent of RMBN has no plausible answer to this question.

But I think that we already have the basis for a reply. Consider first a case of multiple-member audiences. There are at least two variants of this case. In the first, the speaker aims to be addressing a particular epistemic group; in the second, she doesn't. The first variant can be treated as a special case of the situation discussed at the end of the previous subsection: if the salience of the group she meant to be addressing is great enough, so that it was reasonable to form the underwriting beliefs she did, and reasonable to expect that members of her audience would discern this, then the corresponding standard is in play; and if not, then the case can be treated as above (so that there may be no fact of the matter regarding the prevailing standard). The second variant can be assimilated into a case of no substantial mutual belief—the standard will be the 'default' high standard. Note that this gives us a sense of how to handle the other sort of case—where the speaker herself identifies with several groups. Again, if one identification is salient, then this group's standards prevail; if not, there may be no fact of the matter regarding the prevailing standard; or, if it is clear that she aimed to be speaking to everyone, the standard will be the 'default' high standard.[16]

To be sure, these accounts would need to be developed in greater detail; my point here is only that nothing in this sort of objection presents an in-principle difficulty for the proponent of MBN.

4.3. I turn, finally, to an objection which, if successful, *does* present an in-principle difficulty for the proponent of RMBN. This objection focuses on RMBN's implications for what we might call 'epistemically degenerate groups' (such as conspiracy-theory groups of all sorts). If RMBN is correct, the objection holds, then there are cases in which assertions are proper on virtually no grounds whatsoever. To see this, consider the possibility of a group of conspiracy theorists who have (underwriting beliefs generating) undemanding epistemic expectations of one another. Call this group the Conspirators. Imagine that the members of the Conspirators have all sorts of mutual beliefs about what counts as adequate evidence, when conspiracy hypotheses are warranted, and so forth (where these standards are, by our lights, very poor); and assume as well that they share (and recognize that they share) any number of assumptions about conspiracies they take to be real. Then if RMBN is true, these mutual beliefs shape assertoric practice among members of the group; and insofar as assertions conform to the standard that their underwriting beliefs pick out, their assertions are proper. But (the objection contends) this seems just wrong: if they make assertions regarding such-and-such a conspiracy, where the evidence behind these assertions is (we think) objectively weak, surely an account of the norm of assertion should return the verdict that these assertions are improper. RMBN appears to lack the resources to deliver this verdict.

In response, I want to say that even if RMBN is guilty as charged, the charge is not as bad as it seems. To see this, we need only be clear about what RMBN sanctions, and what sort of criticisms we can make of the assertoric practices of the Conspirators even by the lights of RMBN. Suppose that RMBN is guilty as charged: when speaking with each other, the Conspirator's assertions about this-or-that conspiracy are proper whenever these assertions meet the group's (by our lights, poor) epistemic standards. Even if this is true, those of us who reject their

---

[16] This point would lead us to predict that assertions made in print, where one cannot be certain of one's audience or the purposes to which they will put the information they come to endorse, will typically answer to the default standard.

epistemic standards are not without recourse to criticizing the group and its assertoric practice (albeit not by way of criticizing the assertions they make to one another).

For one thing, since we don't share the Conspirator's standards, the result is that, even if RMBN is true, the Conspirators' can't appeal to their own standards to render assertions proper when the assertions are made in contexts in which *non*-Conspirators are in the audience. In such cases, assuming that there is minimal mutual belief, RMBN will set the standard at 'default' and so will deliver the verdict that such assertions are improper. This point suggests that the Conspirators' conspiracy assertions are proper only when speaking to one another, not when engaging the world at large. And I, for one, find this a tolerable implication: we might describe this as a case where they have a very non-standard (by our lights, degenerate) assertoric practice in place.

Still, one might wonder whether this is a tolerable implication. The key question is this: why think that (in cases in which they are speaking to one another) we need to accept their standards, rather than impose our own on them? If we impose our own standards on them, we might well acknowledge that they are making assertions, but we will say that these assertions are systematically improper (no matter who the audience is, and no matter what epistemic expectations that audience has). Let us consider, then, what can be said on behalf of an account of their assertoric practice that regards the proper standards (at least in cases of within-group assertion) as set by what Conspirators themselves mutually believe in context. Such an account squares with the fact that the expectations they have of one another are deeply out of kilter with our own. True, one might recoil at the idea that their within-group assertions are proper. But—and this is the second point I wish to make in response to the objection—we can still criticize their standards, and with it the assertoric practice governed by those standards, *even after we acknowledge that the practice has its own internal coherence* (when the practice is restricted to members of that community).

This is the second sort of way we might criticize the Conspirators' assertions: we might criticize their standards themselves. We can do so even as we allow that, by the lights of those standards, their assertions pass muster. To be sure, we will regard their assertions as improper. But recall that this point is compatible with RMBN: it is only *within-group* assertions that RMBN must regard as proper (when they meet the Conspirators' standards). So long as we are thinking of their assertions as assertions made *to us*, that is, to non-Conspirators who don't share their epistemic standards, their assertions are improper. It is only when we are describing the practice of their making assertions to one another that RMBN delivers the verdict that their assertions are proper. Of course, even this might be too much for some. But consider how we would describe the situation if we assume that knowledge (or some other demanding but invariant standard) is the norm of assertion. We will say that their assertions are systematically improper (a plus); but we will still need to explain the relative stability of their practice, the fact that they all appear to have calibrated their standards with one another, that they have very similar propriety judgments regarding their assertions. To be sure, the explanation is near to hand: Conspirators have many (implicit) mutual beliefs regarding prevailing standards; only (the knowledge-norm proponent will add) those beliefs are *systematically false*. Here I only note that, once it is an open question whether there is one standard governing any assertion made at any time and in any place, this construal of their practice is no longer the only one. Nor is it a particularly charitable one. It seems decidedly more charitable to explain the stability and systematic nature of their practice as a matter of their embracing different standards, and then to register our disapproval of those standards themselves. At the very least, such a position is far less objectionable than we might have thought on first confronting this objection.

This last point is reinforced by a third sort of criticism one might level at the Conspirators, even after one acknowledges that their conspiracy assertions to one another are (sometimes) proper. There is a distinction to be drawn between an assertion itself (an instantiation of a speech-act type) and the proposition asserted on a given occasion. RMBN bears on the question regarding the propriety of the former; it says nothing of the warrantedness of the latter. But for this very reason, we might direct our attention to the propositions that the Conspirators assert, and we can criticize the basis for believing that these propositions are true. We can do so even if, as proponents of RMBN, we allow that their within-group assertions are proper. Of course, this is a version of criticizing their epistemic standards: we reject their standards (and so might reject their assertoric practices, informed as they are by those standards). But it is one thing to say that we reject their standards, another to say that there is no assertoric practice correctly describable in terms of those standards. The objection requires the latter, stronger claim. In light of the facts, first, that this stronger claim in effect is committed to what I called above the uncharitable construal of their practice, and second, that even after we reject this stronger claim we can still criticize the Conspirators in all sorts of ways, I submit that it is best, all things considered, to regard the Conspirators' assertoric practice as RMBN would. I concede that there are costs to doing so; but these costs are, all things considered, outweighed by the other benefits of RMBN.

4.4. I conclude, then, that while the objections suggest the need to replace MBN with RMBN, there is no objection which undermines RMBN itself. This said, I acknowledge that the objections considered here, and in particular the objection from epistemically degenerate groups, do make clear that RMBN does have some striking implications. Some of these might even be properly regarded as a cost that will have to be borne if we accept RMBN. But given what RMBN has going for it, and given that these costs are not as great as one might have initially feared, the case for RMBN remains strong.

5. Conclusion

I believe that there are several lessons to draw from the foregoing reflections. First, even if assertions continue to be made in the face of systematic disagreement, and even if under these circumstances there can be little hope for knowledge or even doxastically justified belief, this does not establish that such assertions are systematically improper. This is because the norm of assertion may well reflect the standing expectations that hearers have of speakers, and that speakers have of hearers, when there is massive mutual belief between them in a given speech exchange. Such a context-sensitive account of the standard fixed by the norm of assertion receives some independent support from a broadly Gricean approach to speech. Such an approach, together with the hypothesis that (academically trained) philosophers form an epistemic group whose epistemic expectations of one another are highly calibrated, enables us to see how philosophical assertions can be proper even when made under conditions of systematic disagreement.

2

Fault and No-Fault Responsibility for Implicit Prejudice
A Space for Epistemic 'Agent-Regret'

Miranda Fricker

> [I]n the story of one's life there is an authority exercised by what one has done, and not merely by what one has intentionally done.
>
> Bernard Williams[1]

How far, or in what manner, should we hold each other responsible for the inadvertent operation of prejudice in our thinking? In recent work in philosophy and psychology on 'implicit bias' this vexed question has taken on a new form and urgency. The difficulty we face in thinking about the structure of our responsibilities in this regard is a canonical one for philosophy: the puzzlement created by conflicting intuitions. On the one hand, we naturally feel that any kind of prejudice that works against another group is something epistemically (and ethically) blameworthy; yet on the other hand, the whole point about 'implicit' biases, including those biases we would consider prejudices, is that their influence in our judgements is so hard to detect that correcting them would seem to require supererogatory, even superhuman, levels of perceptiveness, corrective know-how, or plain time and effort. We are rightly reluctant to proclaim that we are always blameworthy for this kind of epistemic error, for that would surely be an unfair demand in cases where we are structurally[2] in no position to tell that we are guilty of it (a form of non-culpable ignorance), and/or where it would take heroic efforts to reliably correct it (a mere failure to perform the supererogatory).

---

[1] Williams (1993), 69.
[2] By 'structural' ignorance I mean ignorance caused by historical-cultural circumstances, as opposed to personal failure; I develop this idea in Fricker (2012).

Yet we certainly should not throw up our hands and declare that since, as things stand, we cannot reasonably be said to help it, nor can we reasonably be held accountable.³

Other things being equal, epistemic fault is culpable. Given that prejudice is an epistemic fault (I shall specify which fault in a moment), someone who falls into prejudicial thinking is a *prima facie* candidate for epistemic blame. And yet when we consider cases of *implicit* prejudice (where the subject is radically unaware of her prejudicial habit), the idea that the epistemic conduct is blameworthy starts to look pointlessly over-demanding. Implicit prejudice looks to be, at least sometimes, beyond blame. Jennifer Saul has made this point in relation to those biases of which we are unaware and/or which we are unable to control (where these inabilities are, I take it, to be conceived structurally, not as mere personal failing but rather as a product of the society we live in).⁴ She adds, further, the pragmatic consideration that, in any case, going in for the expression of blame may be counter-productive. Jules Holroyd, by contrast, has argued that we are blameworthy for such biases at least some of the time, and the implication is that acknowledging culpability is likely to be part of raising standards.⁵ I take these different claims to give expression to equally intelligible but broadly conflicting intuitions. What, then, are we to make of our mode of responsibility in these matters? My answer to this theoretical question will ultimately deliver certain first-order practical imperatives for individuals, but also, and by immediate implication, for collective institutional bodies under whose auspices the individuals concerned may operate.

Even taking the idea of the 'institutional' as broadly as possible to include all corporate bodies, it is of course not the only arena of prejudice and other bias, for these are also played out in purely personal interactions. However, the institutional does preside over most dimensions of our social activity, whether it be education or employment, including commercial activity, charitable activity, regulatory activity, legislation, law enforcement, legal process, political process, social work, medicine, religion, cultural administration, and so on. In all these spheres of social operation individuals judge, deliberate, and act in role as officers or affiliates of institutional bodies of various kinds—sometimes they perform their role as an individual (the administrator, the teacher), sometimes as members of additive or 'summative' groups (the voters, the patients), sometimes as members of 'plural subjects' in Margaret Gilbert's strong collective sense, where there is a joint commitment to operate 'as one' (the board, the government, or

---

³ I shall use the notions of accountability and responsibility as equivalent.
⁴ Saul (2013), 55.   ⁵ Holroyd (2012), 274–306.

indeed any jointly committed 'we'),[6] and sometimes in looser, more easily disbanded groups united *pro tem* by intentional interdependence (the volunteers, the spectators), as we find theorized by Michael Bratman, or Christian List and Philip Pettit.[7] While these views tend to be understood as competing accounts of collective agency construed as a unified phenomenon, I regard each as capturing a different strength of collectivity.[8] And in these various collective capacities they engage in behaviours and decisions that can have major consequences for others, whether it is spontaneously perceiving someone as 'leadership material', or as carrying a weapon; or perhaps deciding whether they have one's vote, or are suitable to adopt a child; or, again, judging whether they have the mettle to stand for public office. How individuals behave, judge, and deliberate in their various institutional roles is most of what goes on socially; so it is no exaggeration, I think, to describe the institutional as the most all-pervasive and consequentially extended arena for implicit prejudice. It is in relation to that enlarged context, then, that I shall be considering our epistemic accountability as regards the inadvertent operation of prejudice in our cognitive conduct.

I shall focus specifically on *epistemic* culpability, because I take it to be prior to the question of *moral* culpability, inasmuch as generally the question whether someone is morally blameworthy for an act or omission crucially depends on the epistemic question of whether there was non-culpable ignorance in play. I shall, however, present a picture of epistemic responsibility that exactly mirrors a certain conception of moral responsibility—a conception according to which the domain of bad things done for which we are morally blameworthy does not exhaust the domain of bad things done for which we are morally responsible. This conception of responsibility is principally due to Bernard Williams, whose earliest explicit presentation of it is in his seminal discussion of moral luck and 'agent-regret'.[9] Our recently increased awareness of the pervasive influence of implicit prejudice and other biases in our everyday judgements needs fitting into a suitable conception of epistemic responsibility. That is the task of this chapter, and my contention will be that we need a conception that finds a space for *no-fault epistemic responsibility* for certain kinds of bad judgement—a space, in effect, for a first-personal reflexive attitude of *epistemic agent-regret*.

---

[6] For a fuller specification of institutional bodies considered as plural subjects, see Fricker (2010). For the classic early statement of Gilbert's view, see her (1989) or, more recently, (2000).

[7] See e.g. Bratman (1999); and List and Pettit (2011).

[8] For an independent account which orchestrates a complex range of different gradations of collectivity, see Raimo Tuomela (2013).

[9] Williams (1982). See also the discussion of Oedipus in Williams (1993). We find a similarly broad conception of responsibility in the work of Raimond Gaita (see e.g. Gaita 1998, xvii; also Gaita 2011, 86–7).

The possibility of this responsibility status vis-à-vis prejudiced thinking promises to resolve the conflicting intuitions I rehearsed from Saul and Holroyd. We must surely start with the presumption that, at least as regards explicit prejudice, we are epistemically culpable for allowing prejudice into our thinking (though, of course, the question whether it is actually productive to confront each other about it remains another matter).[10] However, my point will be that as we move into 'implicit' territory, where we are likely to find cases where we are not blameworthy for the reason that epistemic bad luck has played an exculpatory hand, *still* we are properly held responsible in the manner of agent-regret. Such *no-fault responsibility* stands to serve as a useful, non-confrontational (because non fault-finding) mode of holding oneself and each other accountable, and thereby pushing for collective ameliorative steps to be taken, as I will try to explain.

The Kind of 'Implicit Bias' that is Implicit Prejudice

Bias is a very general category, which can span many things, from epistemically helpful heuristics to prejudice against stigmatized groups. In order to make my case for the idea that epistemic agent-regret has application, I shall focus on the idea of implicit prejudice and the kind of epistemic fault at stake in implicitly prejudiced thinking. But in order to get the relevant notions in place, let me first borrow a working definition of implicit bias from Holroyd, whose formulation does not explicitly employ the notion of prejudice, but in which the notion of an *automatically* applied *negative property or stereotypic trait* seems at least to incorporate the kinds of negative prejudicial thinking on which I shall be focusing. At the very least, we might think of implicit prejudice as a dominant sub-class of implicit bias as Holroyd defines it:

An individual harbors an implicit bias against some stigmatized group (G), when she has automatic cognitive or affective associations between (her concept of) G and some negative property (P) or stereotypic trait (T), which are accessible and can be operative in influencing judgment and behaviour without the conscious awareness of the agent.[11]

This definition is designed to capture the kind of biases detected not only in Implicit Association Tests (IATs) but also in tests concerning real-world activities such as the assessment of CVs depending, variously, on whether there is a

---

[10] It is reasonable to worry that sometimes interpersonal confrontation will only make things worse; but on the other hand there is evidence that sometimes it can help—see Czopp, Monteith, and Mark (2006), 784–803. I thank Michael Brownstein for directing me to this research.

[11] Holroyd (2012), 275.

male or female name at the top, or, alternatively, a name racialized as black or white.[12] Whatever circumspection one may harbour about split-second differences of click-time on a mouse in pairing up, say, the word 'aggressive' with a black or a white face bearing an aggressive expression, the data about CV assessment and similar experiments involving real-world activities are worryingly impressive.

These kinds of evaluation activities structure many professional worlds—at the minimum they help determine who gets to enter the line of work in the first place, and subsequently who gets to advance in it—so that these sorts of judgements have an enormous influence in shaping the profile of productive social activity over the long term. The implication is that individuals who do not have prejudiced explicit attitudes may nonetheless normally have prejudiced implicit attitudes that have an undetectable deleterious influence on their judgements and deliberations. We always knew prejudice could be stealthy, but putting this new level of undetectability together with the sheer prevalence of the phenomenon produces a whole new perspective. We now appear naively alienated from our own judgements, so that ordinary cognitive self-discipline seems more elusive than ever.

This new image of ourselves, refracted through the lens of controlled experiment, seriously compromises our conception of ourselves as cognitively authentic, or even epistemically responsible. My explicit attitudes are on the whole mine (even when they are borrowed, it is I who have borrowed them); but my implicit attitudes, it seems, are not like that. Inasmuch as they flow from me in a temporally and counter-factually stable manner, I can hardly deny them as a robust facet of my epistemic character; and yet for many of them I would disown their content utterly. This newly alienated self-image is at least as much of a shock to the system as was, in the twentieth century, the Freudian picture of human beings as pulled about by radically unconscious desires and fears. At least the demons of the unconscious were definitively personal to the individual psyche, their quirky forms shaped by our most intimate relations. By contrast, our implicit prejudices against stigmatized groups are peculiarly impersonal in their aetiology, for they are, on the whole, unwittingly absorbed from outside the spheres of intimacy—the attitudinal fall-out from a semi-toxic social environment.

How does Holroyd's working definition of implicit bias effectively incorporate the notion of prejudicial thinking? Let me focus on the idea that someone exhibits

---

[12] Bertrand and Sendhil Mullaainathan (2004) and Moss-Racusin *et al.* (2012). Both these experiments, and many other aspects of the issue of self-regulation, are discussed in Gendler (2014). See also the reply by Nagel (2014). There is also helpful discussion in Holroyd (2012).

implicit bias when she *automatically associates* some *negative property or stereotypic trait* with a *stigmatized group*. Strictly speaking, such an automatic association need not quite amount to a prejudice, but it will in fact do so wherever the association is the result of any motivated failure to properly gear one's attitudes to the evidence—and this will surely be true for most automatic associations of negative properties or traits with stigmatized groups.[13] An evidential shortcoming of this sort might take various forms. It might be a matter of the subject's being guilty of some motivated resistance to counter-evidence, where 'resistance' might be a refusal even to recognize the instance as counter-evidence ('just the exception that proves the rule'), or alternatively it might consist in a failure to follow through with the requisite adjustments elsewhere in one's belief system or deliberations. Or again, the evidential shortcoming might equally manifest itself in someone's being motivated to generalize from an excessively small, or unrepresentative, or contextually inappropriate sample.[14]

Let us gloss these different forms of evidential failing by saying that an attitude is prejudiced insofar as it is the product of (some significant degree of) *motivated maladjustment to the evidence*. A paradigm example of prejudiced thinking would be someone who has the prejudicial attitude that members of social group X are inferior to his own social group, where a significant part of the explanation why he has this attitude is some, perhaps entirely non-conscious, desire for superiority, fear of inadequacy, or perhaps simply a baseline motive to fit in with in-group attitudes. Obviously, prejudice will often manifest itself by way of stereotyping, as is explicit in Holroyd's definition of implicit bias. Transferring our definition of prejudice, we can say that a prejudicial stereotype is a stereotype that is the product of some motivated maladjustment to the evidence.

Stereotyping can itself take different forms. It might take the form of an implicit generalization ('all/most/many Xs are F'); or, alternatively, it might take the form of a 'generic' ('Xs are F') where there need be no pretension to there being a statistically significant number of instances, but merely a bald association expressing the salience of some feature—for instance, what Sara-Jane Leslie calls a 'striking property' generalization where 'striking' indicates danger or

---

[13] Perhaps an exception might be dredged up for purposes of conceptual argument: a case where someone automatically associates a negative property with a stigmatized group *without* any motivation—sheer spontaneous habit of some allegedly non-motivated kind. Insofar as this is a possibility, it inserts a wedge between my notion of prejudice and Holroyd's notion of implicit bias, but a very thin wedge.

[14] Ishani Maitra makes this point (Maitra 2010, 206–7).

risk of harm. (One of her examples is 'mosquitoes carry the West Nile virus', where in fact less than 1 per cent of mosquitoes carry the West Nile virus.)[15]

For many generics, such as the one just cited, we regard them as true. As Leslie convincingly explains, this is because it is reasonable to essentialize certain natural kinds, that is, to regard them as possessing an underlying shared nature, so that properties observed in just a very few instances may reasonably be presumed to flow from that underlying nature and so generalize to the kind as a whole. In such cases the generic habit is epistemically justified, because it tends to lead to true belief. But we can see in what fundamental respect it becomes unjustified when it is applied to social groups, for such groups are thereby treated as if they had an underlying shared essence when in fact they don't ('Muslims are terrorists' is cited as a post-9/11 example of a false generic statement of this kind—understood specifically as an essentializing overgeneralization from a minute sample). This distinctively essentializing type of overgeneralization, fueled by fear of harm of some kind, puts on display its own distinctive form of motivated maladjustment to the evidence, and so comfortably fits the way we are conceiving of prejudicial stereotyping in general.

Having explained how Holroyd's definition of implicit bias (as including an automatic, sometimes stereotypical, negative association) might typically embody the motivated maladjustment to evidence that constitutes prejudice, let me now turn to the question of blameworthiness. I have said that the epistemic fault for which the prejudiced thinker is culpable, other things equal, is the motivated maladjustment to the evidence. Epistemic faults operating unimpeded in the individual's epistemic system are canonical cases of blameworthy epistemic conduct (other such faults might be jumping to conclusions, carelessly overlooking counter-evidence, wishful thinking, dogmatism, sloppy calculation, and so on). If, for example, someone chairing an academic appointments process were systematically, albeit unwittingly, to assess the male candidates' writing samples more highly than those of the female candidates owing to the operation of implicit prejudice in her patterns of judgement, then, other things being equal, we would regard her as epistemically at fault, and so blameworthy.[16]

We might not blame her very much, of course, if she were doing her well-intentioned best under difficult circumstances—pressure of time, lack of institutional support for alternative methods. These are mitigating circumstances, or

---

[15] Sara-Jane Leslie (forthcoming).
[16] I have argued elsewhere that the notion of someone's being blameworthy is best understood in terms of their being *at fault*, and the notion of blaming someone as a matter of *finding fault*, where paradigmatically (though not necessarily) that judgement will be interpersonally communicated to the wrongdoer with a view to inspiring remorse. See Fricker (2014).

excuses, and they function to reduce the degree of appropriate blame, even to zero in some cases, if we accept the possibility of fully exculpatory excuses, which we surely may. But they do not change the kind of epistemic fault that has expressed itself, which is blameworthy, other things beings equal. This is no less so if she herself would be horrified were this to be revealed to her after the fact (as it might be if, for instance, we imagine her as a participant in a controlled psychological experiment on the operation of implicit bias). Indeed, the appropriateness of blame in a case like this is actively supported by the thought that not only might the well-intentioned agent blame herself, but moreover anyone whose work had received a prejudiced assessment would surely blame her too, not only where the prejudice lowered the estimation of the work (a case of testimonial injustice[17]), but surely also where it raised it.[18]

To bolster this idea that our judgements of culpability are partly organized around whether we judge the fault to be traced to, or located in, the subject's epistemic system as opposed to being traced to someone else's (or indeed to the collective at large), let us consider a contrast case. Imagine a situation in which someone justifiedly believes the word of a speaker who confidently tells them that *p*, but who has been culpably careless with the evidence. Our hearer ends up with a false belief, but the story of epistemic fault is such that the buck is passed, and we regard the error as exclusively the fault of the original speaker—she, after all, was the one who had been careless with the evidence, whereas our hearer made no error of reasoning in believing her. The fault is not located in the hearer at all, but rather traced to the speaker. The question 'Whose fault is it?'—heard as a question about the fault's explanatorily salient location—is a very powerful organizing idea in how we make judgements of blameworthiness.

Returning to our present-day implicitly prejudiced assessor of writing samples, we will regard her as epistemically culpable (mitigating circumstances notwithstanding) insofar as we regard *her* epistemic system or character as the explanatorily salient source of the fault. This is so even if we imagine her to be completely unaware of having these implicit prejudices, rather as an implicitly selfish person's selfishness might systematically lead her to act selfishly even while she remains entirely unaware of this fact and (as we often say) 'cannot help it'. This idea of an implicit trait of moral character, especially a vice, makes an instructive comparison—the idea of selfishness often being non-conscious and inaccessible

---

[17] 'Testimonial injustice' happens when prejudice deflates the credibility attributed to someone's word (see Fricker 2007, ch. 1).

[18] I have argued elsewhere that to blame is to find fault regardless of whether one benefits from the fault, so that one can blame even if one does not desire that things had gone differently (Fricker 2014).

through introspection, yet unwittingly manifested in judgement and action, is hardly an alien one. And nor is it (here is the point of the comparison) an alien idea to consider it nonetheless blameworthy—indeed, one might take implicit selfishness as a prime case of the morally blameworthy.[19] The more general point to be extracted here is that certain kinds of moral theory tend to be forgetful of the fact that we normally—canonically, even—hold each other as blameworthy for behaviour that is expressive of bad traits or motives that are understood to be beyond our ken and control. People blame us for any faulty traits or motives they consider *ours*. Furthermore, the point is not restricted to that which is characteristic of the agent, for uncharacteristic motives and acts can still be ours in the relevant sense—features of *our epistemic system*. So whatever reason we might establish for regarding some operations of implicit prejudice as non-culpable will clearly need to invoke grounds other than that we did not know about them, or could not control them, or that they were not really expressive of our epistemic character. All these things can be true of a moment of bias, just as they may be true of a moment of selfishness, and yet I remain a perfectly proper object of blame, excuses notwithstanding, for the simple reason that the fault was mine.

It will be worth an aside at this point, I think, to emphasize the extent to which this observation about a perfectly normal and proper mode of blame is at odds with a philosophical dogma about responsibility. In a range of philosophical debates the idea has become deeply entrenched that we are only genuinely responsible for that which is under our control. Now of course this idea does have application—*some* sorts of loss of control (such as real internal or external compulsion) clearly switch off responsibility altogether—but the idea tends to spread beyond its proper remit to figure as a quite general condition on responsibility, presenting itself as what George Sher has called 'the searchlight view'. According to the searchlight view, responsibility is limited to acts or omissions that one has chosen, or whose causal origins one has chosen.[20] The searchlight view certainly cannot be right, not least because it would rule out cases of culpable neglect and other kinds of culpable ignorance. (It would also rule out bad moral luck of course, and Sher considers a number of cases of that kind, though his discussion is not focused on the concept of moral luck *per se*.) The view he favours, and with which I agree in its broad outline, is one according to which we are responsible for a good deal more than the searchlight view would

---

[19] See Adams (1985). Of the self-righteous person, for instance, Adams says: 'According to the doctrine that all sin must be voluntary, it would seem that he is not to blame. And yet I think he clearly is to blame, not because of his voluntary choices but because of his self-righteous attitude. We will have a lop-sided view of the guilt in human relationships if we do not recognize this' (p. 6).
[20] See Sher (2009).

allow, for we are responsible not only for conduct based on things we know but also for conduct based on things we *should have known* but didn't.[21] To this I would add, in the dimension of control, that we are not only responsible for those things we can control, but also for those things we ought to be able to control but can't (one's profound selfishness, for instance). Obviously any judgements about what someone should have known must be made in relation to some reasonable standard, as Sher quite rightly says. And again, I would add that any judgements about what someone should have been able to control must equally be made in relation to some reasonable standard. Indeed, part of what all this suggests is that there is no escape, in judgements of culpability, from the demands of substantive normative thinking about what is a reasonable demand under the circumstances—there is no available recourse to a criterion of knowledge, or of control, that can guide our normative judgements from the outside.

This aside made, the foregoing reflections on the importance of the explanatorily salient location of the epistemic fault (in me? in her? in society as a whole?) raises the question whether there might be cases of implicit bias where the epistemic fault is *not* located in the prejudiced subject, but merely flows through her. Might there be such cases? If so, perhaps these are candidates for prejudiced thinking that is not epistemically blameworthy. In order to explore this possibility, let us now imagine a different writing-sample assessor. This time she is not engaged in any motivated maladjustment to the evidence *herself*, but merely passively and *innocently* (i.e. without epistemic fault—this is important) inherits the gender biases in play from her environment. Perhaps we can stipulate that she simply has no relevant motivation for the bias; but in addition, we must imagine a situation (no doubt far from that of the author or reader of this chapter) in which she has no reason to suspect that she may be a conduit for gender bias (otherwise naivety on this score would already constitute epistemic fault). In sum, we are imagining a case in which what is going on is the epistemically innocent inheritance of bad epistemic goods from the environment.

Such an imagined case looks like one of epistemically innocent error: our assessor is a faultless (and so blameless) conduit for prejudice. Her judgements are epistemically bad, but it is not her fault. The fault—the motivated

---

[21] Sher, however, believes that something *additional* is needed, namely the condition that the agent's ignorance is 'explained by certain aspects of their constitutive psychology' (2009, 93). Sher considers this necessary to guarantee that an agent's ignorance really is *theirs* where this is to involve the rather strong requirement that the ignorance is caused by something in their constitutive identity. But I do not see the need for this. If we are already talking, as all sides must agree we are, about an act or omission *on the part of the agent*, and of ignorance (culpable or non-culpable) *on the part of the agent*, then that is already enough.

maladjustment to the evidence—has already been committed off-stage by others, the epistemic collective of which she, through no fault of her own, is a member, and whose toxic influence (we have stipulated) she has no reason to suspect. We can ratchet up our subject's epistemic innocence all the more if we imagine her as justifiedly believing herself to be free of gender prejudice—perhaps she is an epistemically conscientious feminist actively trying her best not to let the gender of the writers affect her assessment of the work in any way. If so, this justified (if false) belief about her own lack of bias would constitute *counter-evidence* to the possibility (barely countenanced) that her judgements might involve gender bias ('No risk of that with me—I'm a feminist').

What should we make of such an imagined scenario in terms of our question about epistemic culpability for prejudiced thinking? Given how we have constructed the case, this particular assessor of writing samples is making a faulty judgement, yet the fault is not in her but rather in the collective. She is, of course, herself a member of said prejudiced collective, but mere membership is not enough to implicate her as individually blameworthy, especially given our stipulation that she has no reason to suspect she risks inheriting collective gender prejudice from it. Now, if we assume that epistemic responsibility mirrors the common, narrow view of moral responsibility, to the effect that the only mode of epistemic responsibility for faulty judgement is epistemic blameworthiness, then we are stuck—and our innocent conduit of prejudiced judgements is simply off the hook. We find no fault in her, and the narrow view affords no residual space of responsibility once the possibility of culpability is eliminated. We shrug our shoulders and leave things as they are, for on this picture no epistemic obligations accrue to her in virtue of her serving as the innocent conduit of bad epistemic materials.

But that is a thoroughly unattractive, because conservative, conclusion to draw; and it does not explain the shame she would rightly feel if her prejudicial bias were revealed to her, or indeed if any of those whom she had disadvantaged were to confront her. Is there not something more subtle we can say about her moral status than that she made bad judgements through no fault of her own and so cannot, should not, be held accountable? There is. We can develop an analogy from an alternative conception of moral responsibility, which pictures the domain of moral responsibility for bad conduct as extending beyond the domain of the blameworthy. For this conception we can look to Bernard Williams's idea of 'agent-regret'.[22]

---

[22] See Williams (1982).

Agent-Regret: From Ethical to Epistemic

Williams famously argued that the traditional picture of moral responsibility as coextensive with potential blame delivered a falsely purified version of our normal, nuanced forms of everyday moral consciousness. Furthermore, this purified picture was the imprimatur of 'the morality system'—an absolutist moral outlook which remains as familiar as it is peculiar on account of the fact that it is ('incoherently', as he remarks) part of the moral consciousness of all of us. Were it not for its purifying influence, the natural place of various kinds of luck (in particular, bad luck) within moral life would be readily acknowledged. (Williams registered the peculiarity of the purified conception by reserving the word 'moral' for its special constructive purposes, and instead using the word 'ethical' for a conception of moral life from which the possibility of luck affecting one's moral status had not been theoretically expunged in advance. But I prefer not to hand over a word as resonant as *morality* to the advocates of any conception deemed false, and so will not imitate his linguistic innovation here.)

The specific fruit of Williams's approach to the place of moral luck in our lives, and in particular its implications for how we make sense of cases where people do bad things, perhaps terrible things, through no fault of their own, is something I contend not only moral philosophy but also epistemology can learn from. He argued that when we do bad things through no fault of our own (when we cause bad things through our agency sufficiently proximally to count as *having done it*, yet blamelessly), the natural and proper response is to morally *own* these aspects of our conduct. Williams did not put it in terms of 'owning', but I think it is a highly pertinent psychological trope, and not one exclusive to psychotherapeutic concerns. The schoolroom notion of 'owning up' to having done something bad is simply one of responding to the question 'Who did this?' with the admission '*I* did'. Significantly, such an admission does not entail culpability ('*I* did; but it wasn't my fault...').

This regretful owning of harm done expresses itself in a distinctively agential first-personal reflexive attitude, quite different in kind from that of a bystander, who after all has nothing to own. And it was this distinctively agential regretful expression of responsibility that Williams labelled 'agent-regret'. That agent-regret is a genuinely *moral* response—that is, an expression of moral responsibility and not simply an expression of shock or compassion of the clearly non-moral regret that might properly be felt by the bystander—is revealed by the fact that, typically, there will be moral reasons that apply to the agent which do not apply to anyone else, reasons pertaining indeed to the owning of harm done.

In the example that Williams uses to introduce the phenomenon, the agent is a lorry-driver who, through no fault of his own, tragically runs over a child who has stepped into the road. The response on the part of the imagined lorry-driver is one of horror at what has happened while he was at the wheel, and of a (perhaps impossible) wish to make amends. Significantly, in less tragic cases the agent *may* often be able to make amends, perhaps by offering some kind of compensation; but in a case like the imagined one perhaps the most our driver can achieve is some kind of symbolic expression of deep sorrow at his part in the tragedy. Agent-regret, in my view, is properly considered a feeling of guilt for harm done, though clearly not of a kind entailing culpability (which is why many would recoil from conceiving agent-regret as a form of guilt). Since the morality system has no word, indeed no conceptual space, for any kind of guilt-feeling for non-culpable harm done, where that is considered as a properly *moral* response as opposed to an understandably intensified form of bystander-regret, Williams coins a new term for us.[23]

The lorry-driver case is one of 'outcome' luck: through sheer bad luck the lorry-driver's entirely responsible driving turns out to cause, on this occasion, a tragic effect. Our epistemological concern with responsibilities relating to prejudiced thinking, however, does not relate to unlucky outcomes so much as unlucky inputs, and so it will be useful to present a slightly different sort of moral case in relation to which we may draw our envisaged parallel. Our epistemological concern is with a case in which an epistemic subject has blamelessly inherited bad epistemic goods from her environment (the implicit prejudice), so that the epistemic fault of motivated maladjustment to the evidence has already been committed off-stage by the collective (and we have stipulated that the subject herself has neither been culpably negligent in failing to realize this, nor does her membership of said collective sufficiently implicate her as an individual in this patch of its bad epistemic conduct). This represents what we might think of as a kind of *environmental epistemic bad luck*; one which, furthermore, *obscures from her view the epistemic significance* of the patterns of judgement in which she is engaged.

Let me, therefore, shift to a different example of moral bad luck—one that provides a better fit. Let us consider Oedipus—another case discussed by Williams,

---

[23] Elianna Fetterolf has argued persuasively that a careful reading of Williams reveals an intriguing possibility implicit in his view, namely that, were our ethical thinking to free itself from the undue constraints of the morality system, there would be no continued need for the special term 'agent-regret', because the ordinary notion of remorse might at last be seen to apply to such unlucky cases, instead of remaining confined to cases of fault (Fetterolf 2014).

this time in *Shame and Necessity*.[24] The horrifying things Oedipus has non-culpably done are that he has killed his father and married his mother. As Williams puts it: 'The terrible thing that happened to him, through no fault of his own, was that he did those things.'[25] What Oedipus was not in a position to grasp was the moral significance of killing this man or marrying this woman (in this case, for the simple reason that he was not in a position to know who they were). Circumstances conspired to ensure that he could not have been expected to know these things, and so this non-culpable factual ignorance entailed non-culpable moral-epistemic ignorance of the significance of his deeds. It was without fault, then, that Oedipus committed these crimes; and yet upon their discovery he dashed out his own eyes for shame (an act which in itself symbolizes the reflexive capacity of shame—the internalization of the shaming gaze of others—a capacity for which it is Williams's chief purpose in *Shame and Necessity* to orchestrate an extended argument). The morality system can only make sense of Oedipus's moral shame as a quasi-pathological response—an expression of understandable distress, indeed trauma, but not an expression of moral responsibility itself. The morality system aside, however, Oedipus is appropriately regarded as the unfortunate subject of a cruel environmental moral bad luck, inasmuch as features of his environment obscure from his view the grave moral significance of his deeds.

How is Oedipus's moral misfortune a better moral parallel for our imagined writing-sample assessor, pictured as an epistemically innocent conduit for prejudice? First, whereas the lorry-driver's running over the child was a radically non-voluntary action, the deeds of Oedipus were voluntary (at least under a plain description); and so are the assessments made (at least under a similarly plain description) by our reimagined faultless assessor of writing samples: Oedipus fought and killed a particular man, and married a particular woman; our blameless writing-sample assessor read the samples and judged that these were superior to those. Both do voluntary things, the significance of which, through no fault of their own, they do not grasp; and in both cases their failure to grasp it is down to their circumstances or environment. They both suffer a kind of *environmental* bad luck. For Oedipus the primary bad luck is moral, whereas for our writing-sample assessor the bad luck is in the first instance epistemic (the damage to her judgement is prior to any ethical harm caused). For both, agent-regret is in order; and perhaps for both, too, some degree of shame, though that is neither here nor there as regards the present argument.

My purpose in this chapter is to make available the epistemic counterpart of agent-regret, in order to illuminate the moral status of epistemic subjects in the

[24] See Williams (1993), 58–74. [25] Williams (1993), 70.

position of our writing-sample assessor—a blameless conduit of prejudice who should nonetheless be represented as accountable. What I hope the foregoing discussion reveals is that she would appropriately experience her responsibility for what she has done in the mode of *epistemic agent-regret*. This, in turn, indicates how others might hold her to account, for although Williams himself did not develop third-personal implications of the existence of agent-regret, still the implications are inherent in the first-personal form.

Carving out the space for this emotion allows us to do two things. First, it enables us to represent her as *non-culpable and yet responsible* for her prejudiced thinking. Second, it also allows us to see specific *epistemic obligations* as applying to her, through responding to which she may better 'own' what she has done and put some reasonable effort into putting things right. (I emphasize again, even if this is not the situation of either the author or any reader of this chapter, still it is surely the situation of many people inasmuch as many are still not situated so as to be exposed to information concerning the strange implicitness of implicit bias, and its heavily disguised influence on our cognitive behaviour.) The first point—that she continues to be responsible even while she is non-culpable—we have already covered. But what of the second point, that certain epistemic obligations accrue to the epistemically unlucky subject? What might these be?

The direct counterpart to the moral case will be those obligations relating to making amends—fixing things caused by one's faulty judgement. Some of these will themselves be ethical obligations, or at once ethical and epistemic obligations, such as withdrawing the initial prejudiced assessment and getting the work reassessed by independent parties, no doubt under anonymized conditions. But these are only the most immediate obligations that accrue to the faultless conduit of prejudice. The obligation to make amends will naturally extend to more general aspects of the case, so that we might imagine our assessor, insofar as she is epistemically virtuous, taking steps to raise greater awareness of implicit bias, at least within her workplace, with a view to changing certain procedures so that it won't happen again in the same way. In the case of our imagined assessor, a procedural change such as anonymization might have done the trick,[26] and so

---

[26] I should note, however, that even in a relatively straightforward case such as this, there can be unforeseen consequences. Imagine an example where the anonymization conceals information about certain things, such as age, whose general correlation with how long a candidate has been in the profession might make it relevant to the assessment of the quality of the work (if the writing is a little immature or over-ambitious, then perhaps that is not such a bad sign in someone who is just starting out, but a clearly negative sign in someone who has been writing for several years). Here the seemingly neutral measure of anonymization would influence the assessment of the work in a way that was *less* sensitive to relevant factors, and it might moreover work against the minority or group to which one is especially concerned to ensure fairness—for instance, if members of that group were

the idea of an obligation to push institutionally for greater anonymization is a good candidate for an epistemic obligation rising out of her responsibility status of epistemic agent-regret. The specific epistemic obligations incurred will no doubt often be mirrored by equivalent ethical ones, and will vary from context to context, but they are naturally thought of as essentially compensatory in nature—efforts to make amends and improve the situation in some appropriate measure. As Susan Wolf has argued, there is no algorithm for this kind of thing, and furthermore, the reasons or obligations will tend to be substantially indeterminate; but some kind of stepping up or 'taking responsibility' for what one has blamelessly caused calls on a virtue without a name (Wolf 2013).[27]

The sorts of ameliorative epistemic obligations incurred by the blameless conduit of prejudice, then, will concern the taking of steps to minimize the influence of prejudice in similar processes—perhaps institutional measures, such as the removal of names at the top of writing samples for long-listing purposes in appointment processes. But now we encounter an important feature of the case: the individual's power to ameliorate the situation is very limited. The kinds of procedural change we are imagining might help make amends and ensure it doesn't happen again are fundamentally *institutional* changes. These are changes that individuals can push for in the institutional context (the administrator, the teacher), but if they are to be implemented, then a more collective effort is called for, whether on the part of a loose association of

---

more likely to be early career candidates. Of course, ideally we would be able to learn from such a case that the appropriate form of anonymization in this context is one where names and institutions are removed, but not age, or not 'years since PhD', or whatever. But this learning process takes time, and in any case it is extremely hard to detect when assessments have indeed been skewed. I thank Jo Wolff for drawing my attention to this sort of possibility.

[27] David Enoch has further developed the theme of moral luck cases involving an obligation to 'take responsibility', which he regards as expressing the grain of truth in an otherwise 'seriously flawed' discussion on Williams's part. In short, Enoch rejects the idea of moral luck, and tries to explain it away with the idea that whenever one is in a situation of purported moral luck, what's really going on is that one has incurred a moral obligation to take responsibility after the fact, so that if you fail to honour that obligation, you are at fault. Thus we find ourselves back in the familiar confines of morality narrowly construed—a conception according to which moral responsibility can express itself only in success or failure vis-à-vis the demands of moral obligation. Enoch rejects the idea that the unlucky agent is *already* responsible—tragically responsible, for instance, for the death of a child who ran into the road while one was at the wheel. While I agree that failure to respond to incurred obligations to make amends would indeed typically be a matter of fault, insofar as Enoch's purpose is to *replace* Williams's claims about agent-regret with claims about 'taking responsibility' I regard it not as rescuing any isolated insight from Williams, but rather as ultimately contorting the central insight back into the straightened form Williams was aiming to leave behind. I thank David Enoch for helpful discussion of these points, over which we agreed to disagree when I presented this material at the Carlsberg Institute in Copenhagen.

colleagues and/or on the part of a committee or board which might be more tightly knitted together by joint commitment and so operate as a plural subject.

The fact that individuals are unlikely to be able to do very much acting alone exposes the fact that the fulfilment of individual obligations in the institutional context will tend to require agitations that inspire more collective responsiveness to the ameliorative obligation. Such agitations will often be sufficient to generate new obligations on the part of relevant collective bodies in the institution, largely because raised awareness of failure tends to generate an obligation to improve where possible. And improvements *will* often be possible. For instance, our biased but blameless writing-sample assessor might press for anonymization of writing samples. If, in doing so, she makes the appointing committee as a whole aware of the problem, with or without a concrete proposal to improve things, then this will generally be enough to generate a collective obligation on the part of the committee. Given that the individual is already operating in some institutional role (she assesses the writing samples as part of her job, after all), it may be that the obligations she incurs in that capacity *already* involve the institution under whose auspices she is operating. But the point here is that even when that is not the case, her raising awareness in the relevant collective body will tend to be sufficient to generate a new collective obligation to help improve the situation, whether through retrospective compensatory measures or forward-looking ameliorative steps, or both. Such collectives are thereby obliged to take responsibility after the fact.

This is significant because the lesson we should take from our example is that, when it comes to implicit prejudice, the individual is rarely the only epistemic agent that acquires the epistemic obligations arising from an individual's epistemic agent-regret. There will tend to be immediate collective normative repercussions. Ultimately, then, the take-home message here is not only that there is a space for epistemic agent-regret, but that the epistemic obligations generated by that zone of no-fault responsibility are not confined to the individual who suffers the agent-regret, but extend swiftly to relevant collective bodies too.

Let me finish by looking a little more closely at what specific form the ameliorative obligations might take: what sorts of things would in fact compensate, mitigate, or thoroughly pre-empt the operation of implicit prejudice in our practices such as assessment and selection in a competitive process? A significant part of what is needed is for institutional bodies to pursue formal procedural techniques that minimize situations in which bias can enter in to influence proceedings (for example, through greater anonymization). Some, however, may make an additional case for the institutional provision of de-biasing therapies of the kind being researched in psychology. It is already standard in universities, for instance, for staff to be required to attend some kind of equal-opportunities-awareness training prior to serving on an appointments panel. In future one imagines it might be part of such a requirement that one take certain appropriate de-biasing therapies. In recent papers Jules Holroyd (2012), Tamar Gendler (2014), and also Alex Madva[28] have emphasized the opportunities for de-biasing oneself that these kinds of therapies seem to offer, and rightly so. But there is surely a serious proviso here: we have not yet reached a stage of stable confidence in any given set of de-biasing therapies, and perhaps may never reach such a stage. What would be needed, after all, is a set of well-established, cross-contextually effective, and practicable de-biasing techniques whose results are widely known to be reliable in the differing contexts into which they are introduced and across different social types. We are nowhere near that situation yet, and to urge the adoption of such therapies too soon would carry a significant risk of unintended consequences as regards bias, and also of causing resentment and backlash if they do not ultimately inspire trust. While a certain imaginative daring is needed as regards increasing purely procedural mechanisms that eradicate the possibility of bias; caution must surely remain the watchword as regards introducing de-biasing therapies in the workplace, even on an optional basis.[29]

With this cautionary note now struck, there is every reason to keep seeking solutions of all sorts. When we allow prejudice into our judgements and deliberations we are always epistemically responsible, if not culpably, then non-culpably in the mode of epistemic agent-regret. The burden of this chapter has been to explain how, even when individuals are non-blameworthy for implicit prejudice, they are (a) appropriately held responsible; (b) typically acquire certain obligations of an ameliorative sort; and also (c) these obligations quickly extend to any collective body or bodies in the institutional structure under whose auspices they are acting, because raising awareness of a dysfunction in the occupational business of an institutional body is generally sufficient to create collective obligations for that institutional body to take ameliorative steps. This is how a defect of individual epistemic conduct may generate not only individual obligations but also collective obligations to instigate change, for it is essentially through these larger agents that we may work more effectively towards an improved epistemic environment.[30]

---

[28] See Madva (unpublished manuscript).
[29] See the general note of caution struck by Jennifer Nagel regarding the (in)accuracy of certain de-biasing strategies (Nagel 2014, sec. III).
[30] I have given versions of this chapter in a number of places: Carlsberg Institute, Copenhagen; University of Cambridge; University of Stirling; Dartmouth College, New Hampshire; University of Bristol; University of Manchester; University of Cape Town; and Abö Akademi, Finland—I am grateful to all those present on these occasions for helpful discussion.

3

On Knowing What We're Doing Together

Groundless Group Self-Knowledge and Plural Self-Blindness

Hans Bernhard Schmid

Introduction

An influential view argues that in order to act intentionally, the agent needs to know what he or she is doing. Such self-knowledge, it is claimed, is epistemically distinctive in that it is 'groundless'—non-observational and non-inferential. This chapter investigates how this view relates to the theory of intentional joint action. Is our knowledge of what we are doing *together with others*—collectively, as a team or a group—of the same groundless kind? The chapter is divided in three sections. The first section unpacks the idea of groundless (individual) self-knowledge, as developed by G. E. M. Anscombe, and in more recent philosophical research. Plausible features of self-knowledge include first-person identity, first-person perspective, first-person commitment, and first-person authority. The second section plays the part of the individualist's advocate. In order to know what *we* are doing, it is obviously not enough to know what *I* am doing; I need to have an idea of what my partners are up to, too, and it seems that observation and inference are the only sources of knowledge of other people's actions. I might not know exactly what it is what my partners are doing (for example, in a company with a high degree of division of labour), but for me to

---

I wish to thank Miranda Fricker, Michael Brady, Katharina Bernhard, the participants of the Social Ontology Colloquium at the University of Vienna, as well as an anonymous referee for helpful comments.

know what it is *we* are doing, as a company, I need to have *some* knowledge of the others' doing, and it seems that any such knowledge cannot be of the non-observational or groundless kind. The third section argues against the individualist view and defends (a moderate version of) groundless group self-knowledge. The knowledge in question is plural pre-reflective and non-thematic self-awareness of what it is the participants are doing together. Groundless group self-knowledge, however, differs from groundless individual self-knowledge in the kind of identity, perspective, commitment, and authority it involves.

1. Groundless Self-Knowledge Unpacked

The claim that intentional action implies knowledge of what it is one is doing may not initially strike one as plausible. After all, large parts of intentional action are performed unthinkingly—we're not monitoring it actively. Nevertheless, we're acting intentionally even if we're not paying any attention to it. In that ambitious sense, we certainly do not constantly need to 'know' what we are doing. In a more basic sense, however, it seems that even in such cases of routine action we do, and our agency breaks down if we don't. That more basic 'knowledge' involves a sense in which, even in routine behavior, it has to be apparent *to us* what it is we're doing, or else intentional action breaks down.

Such breakdowns sometimes happen. Imagine that during a short break after some hours of intense work on a paper at your desk, still thinking about your paper, you find yourself in the kitchen, opening the fridge, not knowing what it is you're doing there. Perhaps your cluelessness does not run all the way down to your present bodily movements—you know perfectly well that you're opening the fridge—but you have no idea as to the question of why you're doing it. Were you about to get something from there, or put something back? You still feel utterly lost and rather stupid for a moment, and intentional action has broken down, until you realize that you're holding a cup of black coffee in your hand. It is coffee time. And you don't drink your coffee black. So that's what you were about to do: get some milk, and there it is, the milk, right in front of you. You're now able to resume your intentional action—you're back on track. The crisis is over.

A philosophical question is, what exactly has gone wrong here? One potential answer might simply be that you have lost for a moment the very element that brought you back on track at the end of the crisis: you were lost in your thoughts about your paper, failed to watch what you were doing, and you were thus unable to jump in when your auto-pilot somehow failed. This implies a certain view of how intentional action works: if the action is not of the altogether spontaneous sort, there usually is a decision to act that is then kept in mind, and that places constraints on the future behavior which the agent has to observe with sufficient attention. Another view argues, however, that a crucial element is missing in this account. An agent needs to know what it is he or she is doing 'just like that', without resorting to recollection that a decision has been made sometime in the past, and perception of the stage of execution in which he or she is currently engaged. The shortest argument for this latter view is that if somebody else asks you what it is you're doing, you don't *normally* have to observe your behavior in order to come up with an answer. The rare cases in which you do are typically cases in which your intentional action has already broken down, such as in our fridge example.

This is similar to the view argued for by Anscombe in her *Intention*:

Say I go over to the window and open it. Someone who hears me moving calls out: What are you doing making that noise? I reply 'Opening the window.' I have called such a statement knowledge all along; and precisely because in such a case what I say is true—I do open the window; and that means that the window is getting opened by the movements of the body out of whose mouth those words come. But I don't say the words like this: 'Let me see, what is this body bringing about? Ah yes! the opening of the window'. Or even like this 'Let me see, what are my movements bringing about? The opening of the window'. To see this, if it is not already plain, contrast this case with the following one: I open the window and it focuses a spot of light on the wall. Someone who cannot see me but can see the wall, says 'What are you doing making that light come on the wall?' and I say 'Ah yes, it's opening the window that does it', or 'That always happens when one opens that window at midday if the sun is shining.' (Anscombe 1957, 51)

Thus the kind of knowledge Anscombe places at the heart of her account of intentional action is not based on observation and inference. It involves 'opinion' that, as she famously puts it, 'is held without any foundation at all' (Anscombe 1957, 50)—if it is knowledge, it is groundless. Anscombe claims that it is knowledge of this kind that characterizes an action as intentional.

What, then, is this groundless knowledge? Anscombe further characterizes it in contradistinction to the observational knowledge one may have about one's actions:

[I]f there are two *ways* of knowing here, one of which I call knowledge by observation of what takes place, then must there not be two *objects* of knowledge? How can one speak of two different knowledges of *exactly* the same thing? It is not that there are two descriptions of the same thing, both of which are known, as when one knows that something is red and that it is coloured; no, here the description, opening the window, is identical, whether it is known by observation or by its being one's intentional action.
(Anscombe 1957, 51)

Anscombe is very explicit in her rejection of the view that it is only the intention (and nothing about the actual behavior) that is 'known without any foundation', but she says surprisingly little about what this groundless knowledge really is. Reconstructive work needs to be done here, a great deal of which can be found in the relevant literature. In what follows, I'll consider and explain four features that self-knowledge of the relevant kind seems to have:

(a) First-person identity. To use Anscombe's example, the knowledge in question is whatever is lacking when I hear that somebody is ringing the bell, but I do not realize that it is *me* who is ringing the bell—I hear that the doorbell is ringing, but I'm not aware that it's me who's pushing the knob by leaning against the wall. This characterizes my ringing the bell as non-intentional, even though it is me who does it (Anscombe 1957, 51). The way in which I become aware of the fact that it is *me* who is doing the pushing is not by learning it is the person who is leaning against the wall who is doing it, because that does not, in itself, constitute the kind of self-knowledge needed here. I still need to know that *I* am that person, and no observation alone can yield that knowledge. Anscombe's point is closely related to Wittgenstein's distinction between the subject use and the object use of 'I' (Wittgenstein 1958 [1933/4], 66 ff.). The 'groundless knowledge' in question is knowledge of what it is *I* am doing, and it involves self-knowledge that is radically different from the way in which knowledge of what *somebody* is doing is knowledge of that agent. Self-knowledge of the 'groundless' kind does not *refer* to oneself, while observational self-knowledge does. Groundless self-knowledge does not 'pick out a person', to use Wittgenstein's expression. It establishes the agent's identity in a way that is not open to any question concerning the referent of the 'I'. Perhaps it is even wrong to say that groundless self-knowledge *identifies* the agent. If the question 'is it really *me* of whom I'm thinking when I think that I'm doing such-and such?' is nonsensical, this is not because the reference of that use of 'I' is infallible, but rather because it does not refer at all. This kind of self-knowledge is not 'of' the self in that referential way.

(b) First-person perspective. Sidney Shoemaker spells out the unique way in which groundless self-knowledge establishes the agent's identity. He, too, approaches the topic *ex negativo*, with an analysis of what an agent without that kind of self-knowledge would be missing. Such an agent would be 'self-blind' (Shoemaker 1996, part 1). 'A self-blind creature would be one which has the conception of the various mental states, and can entertain the thought that it has this or that belief, desire, intention, etc., but which is unable to become aware of the truth of such a thought except in a third-person way' (Shoemaker 1996, 30 f.). Self-blindness would not simply result in some limitation on the agent's rational capacities, he claims. Rather, it would disqualify the agent in question as a rational altogether. A rational creature is a creature that is capable of reasoning, and reasoning has to start from the reasoner's own attitudes. For a creature that has only perceptual access to mental states, however, there is no way to know of a mental state that it is its own in the way required for reasoning.

Here is a version of the story that is usually used in the literature to illustrate the point (it is the same point as in the Anscombian ringing of the doorbell, only that it is attitude rather than action that is at stake here). Imagine a highly perceptive agent who *knows* what a person's intentions are simply by looking at that person's face. However, looking at faces is the only way she knows about intention. Luckily, she has a mirror, and she has her own face in view. How can she know that whatever attitude she reads off that face is hers? She knows that mirrors usually show the faces of the people standing in front of them. To make sure that it is her who is standing in front of the mirror, she can resort to another mirror that shows herself standing before the first mirror. Now she sees not only the expression in the mirror; she also sees whose expression is reflected in the mirror. But this pushes the question only one level further back: how can she know that the person whose expression is in the first mirror is herself? Further mirrors can't help. 'Self-blindness' cannot be cured by self-observation of whatever high a level, Shoemaker argues, because no such observation provides first-personal access. Thus the attitudes the person observes herself to have are not available to her *as her own* in the way that is necessary to engage in reasoning.

(c) First-personal commitment. A self-blind person fails to see his own attitudes as his commitments, that is, as accepted constraints on his thoughts and actions. This may not seem particularly important, though. Louis C. K. once said about his most cherished beliefs and intentions—such as the belief that other people's well-being matters, and the intention to do good—that he is very proud of the fact that he has them, but that he is certainly not planning on acting on any such attitude. He likes the 'having' part, he says, but that does not amount to any sort of commitment. But insofar as this is true, it seems that these attitudes are not really *his* after all, or not his *in the right way*. He may like to *think* he has them, but he really doesn't. It seems plausible that an attitude that does not come with any sort of corresponding commitment to think and act cannot be *an agent's* attitude.

This is brought to the fore rather forcefully in Shoemaker's discussion of Moore's paradox (cf. Shoemaker 1996, 34 f., 74 ff.). There is nothing *formally* wrong with views such as 'p is true, but I believe it is false', because both conjuncts may very well be true: it may be raining, while I believe that it is not raining. However, the position taken by somebody uttering such a sentence is obviously not viable—and neither is the position taken by somebody having such an attitude, and the point at stake is that such an attitude fails to take the subject's own attitudes as his or her *commitment*. The attitude may be formally sound, however, it contains contradicting commitments. By thinking 'p is true' the subject is committed to the truth of p, but by thinking 'I believe p is false', the subject is committed to the falsity of p.

A similar line is pressed by Shoemaker, who argues that perceptual models of self-knowledge fail to account for the way in which an agent's attitudes are his or her commitments, and for the way in which Moore's paradox is a paradox. A self-blind person with the above mind-reading capacities may look in the mirror and see that she believes that it is not raining, and then look out of the window and believe that it is raining, amounting to the belief that it is raining, and that she does not believe it. Another look in the mirror would of course quickly reveal to the self-blind person that she has changed her opinion—the person in the mirror has changed her views on the weather. But any such move reduces Moore's paradox to a mere case of error—after checking the weather, and before looking into the mirror once more, the person mistakenly believes of herself that she does not believe that it is raining, while she has changed her opinion on the weather at the moment she has looked out of the window. Even if we remove the temporal lag by assuming an inner sense of self-perception that works simultaneously with the 'outer' senses, the gap between what a person believes to be the case and what a person takes herself to believe is not closed *in the right way*—Moore's paradox is not just some sort of mistake, after all, but a paradox.

This carries over to Louis C. K.'s case to the degree to which a similar point can be made for the case of intention. Assume for a moment that intention is always under the guise of some good—whenever we do things intentionally we do so because we see the doing as worthwhile under some description. The practical version of Moore's paradox would then be an assertion or attitude of the sort 'p is utterly bad, but I intend it'. Just as in Moore's cognitive case, that is an attitude that is unproblematic when it refers to other people's intention, but it is not an attitude that can be taken towards one's own intentions.

A self-blind person would not be committed to his or her own views in a way that would make any form of reasoning possible, because he or she would not be bothered by any inconsistency between his or her own attitudes. The third-personal insight that he or she has it wrong cognitively, or that his or her intentions are thoroughly misguided, would amount only to the view that there is *somebody* who got it wrong and who is thoroughly misguided—which is none of *his or her own* business. There is simply no bridge between the insight that *somebody* obviously has a problem to the insight that this problem is *his or hers*.

(d) First-personal authority. A further point to be made for the case of groundless self-knowledge is that it seems to explain the special *authority* agents seem to have over the aspect under which it is intentional. This authority consists in the way in which the agent's view of what it is he or she is doing *settles* the question of under which description it is intentional. This authority may not seem obvious. You think you're calling your friend, but you type in your parents' number. Where is your authority? An outside observer may well be in a better position to know what it is you're doing. Still, you may claim that calling your friend was what you *intended* to do. But even authority over intentionality may not seem obvious. To modify an example discussed by Velleman: You quarrel all evening with your friend, and it is only later on that you realize that you intended to break up that friendship—a fact which you denied when your friend brought it up; another person was obviously in a better condition to say what it is you were doing intentionally. Yet again, any such case is a challenge to the *range* of first-person authority, not to its existence. After all, it was intentionally, and known to you to be so at the time, that you brought up all these sensitive issues in the discussion—you did not just sit there and listen in surprise at these words coming out of your mouth. Thus there was something you did know non-observationally, even though you may have been unaware of, or mistaken about, the reasons why you did what you did. Quarrels over what it is you're doing may arise in the interpretation of the relevant description of your action, but any such dispute does not displace you as an authority over what it is you're doing. Within his discussion of Anscombe's groundless self-knowledge, Richard Moran (2001, 124 ff.) argues that the 'knowledge' in question should be more adequately conceived of as a 'making'. It is no wonder that people should seem to have special epistemic authority over what it is they intend, Moran argues: they are the *authors* of these states after all, that is, the ones who have made up their own minds accordingly. One special feature of Anscombian 'self-knowledge', in other words, is just the fact that it is a product of agency (cf. O'Brien 2003). According to the view that emerges from Moran's (and others') view, first-personal authority is the feature in virtue of which the agent is 'in control' of his or her attitudes (Moran 2001, 147). Thus, first-personal authority marks the agent's place in action. It does not exclude the possibility of a great deal of estrangement, as is the case in weak-willed action. But such estrangement cannot be total, or agency breaks down.

Moran and others have attempted to demystify 'groundless belief' by arguing that it is no wonder that our relation to our own mental states should be special, given we are the 'authors' or makers of our own attitudes by making up our mind (Moran 2001)—a view that seems especially plausible where intention is concerned, because it may appear that whereas we're not free to choose what to believe, intention in terms of decisions what to do is 'up to us' to a somewhat wider degree (cf. Paul 2009. Other authors attack the view that self-knowledge is somehow basic and 'groundless' head-on. Shoemaker has been accused of conflating introspective awareness and self-knowledge, and of implying that a quasi-perceptual 'inner sense'-view of introspection can be maintained, amounting to the view that the self-knowledge in question is perceptual after all, only that the sense in question is of a particular kind (Kind 2003; Armstrong 1981). Looking at the literature, it is far from obvious that the 'Anscombian view' is generally accepted, even if it is as loosely conceived as in the remarks made in this section. Many of the arguments sketched above are controversial. But I think it is plausible to say, at least of most of the competing accounts, that their aim is not to deny first-person identity, first-person perspective, first-person commitment, and first-person authority *as such*, but rather to account for it in a different way—for instance, by relying on introspection rather than on non-observational self-knowledge. This is enough for the line of argument of this chapter to proceed, because it seems that *any* account of self-knowledge that features first-person identity, perspective, commitment, and authority is likely to run into at least some of the problems to be discussed in the next session. The problem is that any knowledge of what it is *we* are doing together involves knowledge of other people's actions. And it seems that any such knowledge needs to be of the observational or inferential kind. You may know what *you* are doing 'just like that'—but not what others are doing. Looking at the matter this way, it does not seem difficult to argue that knowledge of what we are doing together does not come with first-person identity, perspective, commitment, and authority.

2. How Do We Know What We're Doing Together?

Typical cases of joint actions are *intentionally* joint. When we go for a walk together we do not *coincidentally* walk alongside each other. The joint walk is intended by us *as a unit*. The fact that a joint action is one token action with many participants is sometimes downplayed in the literature, and the reason may be that it is felt that this compromises the status of the individual participants as agents. But there is no reason for such worries, as the analogy to complex individual actions makes clear. Complex individual intentional actions often consist of different parts that are themselves intentional actions. Filling the coffee machine with beans and water, switching it on, taking the milk can from the fridge are intentional actions that combine to constitute one token intentional action, the making of the coffee, without thereby losing their status as intentional actions (cf. Anscombe 1957, §§22–3). Analogously, the status of the participating individuals as agents is not undermined if their individual contributions combine to a single token joint action. But it is clear that not just any aggregate of individual intentional actions will do. Just that I happen to fill the coffee machine, you happen to switch it on, and he happens to get the can of milk, together with similarly coincidental other actions, does not mean that there is one token action that is jointly intended. What is it, then, that provides the form of unity that is necessary for an aggregate of individual acts to be components of a joint intentional activity?

At least one attempt has been made in the received literature to give a straightforward Anscombian answer to that question, applying her approach to intentional action to collective action (Laurence 2011). Ben Laurence uses Anscombe's view on how single acts may be parts of one and the same intentional action of making tea as the starting-point. The focus is on the non-accidental nature of the actions, which is explained with the special form of mutual explanatory relations between the overall intentional action of making tea, and the component actions (warming the water, preparing the cups etc.), as well as between those component actions. 'The several actions thus share an explanatory unity' (Laurence 2011, 277), and it is not difficult to show that the same kind of 'non-accidental fit' is also in place between the participant actions of the members of a joint action—if we are intentionally walking together, I walk and you walk because we're walking together, and vice versa. This is the basic line of this 'Anscombian Approach to Collective Action', and it is summarized as follows: 'What emerges from the comparison of individual to collective action is that when a group is acting together, the actions of the individual group members can be straightforwardly instrumentally rationalized by an action being performed by the group' (Laurence 2011, 281); if this 'straightforward instrumental rationalization' is possible, the participants can be said to share one token intention (Laurence 2011, 283). This approach nicely explores the analogy between individual action and joint action, and it does place the group in the role of the agency in a way that does not displace the participating individuals from their role as agents. But this does not so far place the core Anscombian idea in the analysis: what about the particular 'groundless' way of knowing what it is we're doing together? Laurence argues that it exists, even though this may not be true of all members. '*Some* suitably placed person(s) must know without observation what purposes the group is pursuing. A group cannot be said to ϕ intentionally, if none of its members know that it is ϕ-ing, or even if some do but only through observation.' The argument concludes by claiming that group groundless self-knowledge of that sort does not imply any group mind. Collective action is action by a unity of embodied individual minds: 'In light of this, why not say that a collective agent knows what it is doing, and why, without observation.' Laurence is, however, not entirely comfortable with that claim in the end. He emphasizes that 'knowledge without observation' is a purely philosophical term invented by Anscombe for a specific purpose, and he admits that it might be 'laying it on a bit thick to speak in this way' of groups (Laurence 2011, 293). Groundless group self-knowledge, in the view argued for in Laurence's paper, simply amounts to the claim that practical reasoning works in the same formal way in collective action as in individual action.

Let us grant that we reason from a we-perspective if we act together. But as the above discussion has suggested, we need to dig somewhat deeper in order to be justified to call any such we-reasoning 'groundless self-knowledge' of joint intentional action. After all, we have encountered a whole series of features of what groundless self-knowledge seems to imply. Before jumping to the conclusion that the same self-knowledge applies in the case of joint action, we better examine how this works with each of these features. What about first-person identity, first-person access, first-person commitment, and first-person authority in the plural case? It seems that if an account of collective action aspires to be straightforwardly 'Anscombian', these issues have to be addressed.

This is where the problems with the assumption of groundless self-knowledge of joint action start. None of the features encountered in the analysis of groundless self-knowledge carries over smoothly to the case of joint action, as shall be argued below. It seems that whatever first-person identity, first-person perspective, first-person commitment, and first-person authority there is in the *plural*, it works nothing like the singular case. To the degree to which this can be made plausible, it seems that groups cannot *really* act intentionally in a straightforward sense, and that what groups can do is really more like an *emulation* of one token intentional action that individuals can pull off if they combine in a suitable manner. The joint action in question will then not *really* be one token intentional action, but a complex of actions patched together so as to look as token-like as possible. I will make the case against groundless group self-knowledge in this chapter, arguing against the possibility of something like the *singular kind* of first-personal identity, perspective, commitment, and authority in plural case, before turning the tables in the concluding section.

(a) First-Person Plural Identity. The first difference seems rather obvious. Self-knowledge of one's own individual actions may establish the identity of the agent in an immediate way that involves no referential 'picking out' of any kind. This does not seem to work in the plural. As a participant, it appears that you don't know who's with you *in the same way* in which you know you yourself are involved. The only non-referential, immediate, and perhaps doubtless feature of any 'we' is the 'I'. Any attitude of the form '*We* are doing x', '*we* intend to x', '*we* believe that x', is clearly quite heavily involved in the picking-out business, and may well go wrong accordingly. There is no question to whom the 'I' in the 'I'm doing...' refers—because it does not refer, in Anscombe's view, and even if the line is taken that it does refer after all, the reference in question can plausibly be claimed to be of the kind that is not open to doubt. Any plural version of such attitudes, however, does leave ample room for mis-identification, and is thus subject to reasonable doubt. No plural identity is established 'immediately' by somebody having an intention of the plural form. I might think that 'we intend...', but I'd better *know* who's in and who's out, and that knowledge does not come 'just like that'. I may *mis-identify* the members of the relevant group (I mistakenly take you to be one of those of whom I think in terms of 'we intend...'), or there may not be any such identity at all, such as when I'm dreaming that we're doing such-and-such with nobody around. While the self of groundless individual self-knowledge is somehow special with regard to its identity, no plural self that figures in some individual's attitude can aspire to any such privilege.

This carries over to the way in which the knowledge of what we are doing combines to constitute knowledge of a single token action. It seems that joint intentional activity resembles much more the initial fridge case, where I know 'without ground' the 'current bit'—my opening the fridge—but have to resort to recollection and perception as to the question of what it is I'm doing. If we are walking together, I may have whatever epistemically groundless (or perhaps introspective) knowledge of what it is *I* am doing. As far as your part is concerned, however, I'd better recall correctly what exactly it was we've agreed to do together. But that's not enough. Just that I recall our agreement, and thus normatively expect you to do your part, or that I am engaged in some such instrumental or explanatory reasoning, doesn't make it the case that this is in fact what we're doing. I may still be mistaken. Eventually, I'll simply have to watch and see; what I take 'our intention' to be does not settle the question of what it is we're doing together in the same way it does in the case of my own intentions. The question 'how do we know what we are doing?' does not seem to be epistemically on a par whether it is asked in the distributive or in the collective sense. Distributively, each of us may know 'without ground' what he or she is doing; collectively, however, as a team, we do not seem to have any such knowledge.

If you're hiking up a narrow path, and I ask you what it is you're doing, you'll be able to answer, 'I'm climbing that mountain', just like that. If, however, you're climbing that path together with a friend who is walking behind you, and I ask you what you're doing, you'd better turn around and see if he's still with you before answering, 'we're climbing that mountain'. This may not seem much of a difference in principle. The fact that the other is still going along with you—a matter of perceptual knowledge—may seem to matter no more to your climbing the mountain *together* than the existence of the mountain—another matter of observation—matters to your individual climbing venture. Yet an important difference that is at stake here is brought to the fore when intention rather than action is at issue. You may know what it is you *intend*, individually, and perhaps you know it just like that. As to the intentions you take to be jointly held, with others, you'd better be able to quote some evidence for what it is the relevant other people have in mind. One can be painfully mistaken in assuming of an intention that it is 'ours' in a way that does not seem—at first glance at least—to have any correspondence in the individual case. You may mistakenly think that you're sharing an intention in a way that differs from individual intention. Perhaps you misunderstood the other person whom you have met on the trail, and you assume now that the climbing is something you're doing together, because that's what you have agreed to do, while he still thinks of the hiking that is going on just in terms of a coincidental case of one person walking behind another, and not in terms of one token action. In that case, you're wrong at a level that does not seem to have a parallel in the singular case.

(b) First-person plural perspective. The difference is further highlighted if we focus on the 'how do you know?' question. If you tell me what you intend to do, individually, it does not make much sense for me to ask how do you know what it is you intend. You just know—that's it. But if you tell me, that is what you intend to do together with your partner, no such reply seems to be possible. You don't 'just know'. You'll have to quote some evidence, and you are likely to reply with: 'That's what we have agreed to do, and here we are', or some such. In that sense, too, joint intentional activity seems to be deeply different from individual intentional activity. Whatever knowledge of what it is we are doing *together* cannot be basic, but implies individual self-knowledge and observation. Neither of us has immediate awareness or introspective access to *our* intention to go for a walk together, but only to whatever individual contributive intentionality we have, individually. As no opinion an individual might have of whatever she thinks she is doing together with others *guarantees* joint activity, and as our knowledge of what we are doing, in typical cases at least, involves a great deal of observation and inference, there is a sharp contrast between individual and joint action (remember, though, that this section is playing the individualist's advocate, and that a different line of argument will be developed in the next section).

(c) The same applies to the question of first-person plural commitment. If a team or group of agents jointly settle on a view of what is true and what is good, this places the participants under some normative constraints—if 'we' believe that x is true or good, each of us can be normatively expected to treat these attitudes as valid in our reasoning. Whatever normativity there may be in attitudes that are 'ours', however, it is clearly not of the kind of commitment that is involved in one's own intentional states. This comes to the fore in Moore's case. 'P is true, but I believe it is false' is not a viable position; 'p is true, but *we* believe it is false', however, is actually quite frequent in our lives, as anyone who has ever participated in a department meeting knows (for example, your department has settled on a view of how its problems can be solved—you, however, know perfectly well that the planned measures will get your department only deeper into trouble). And it is not conceptually incompatible with the capacity to reason; in fact, it is often by the most reasonable group members that this attitude is taken. This does not only apply to belief, but also to intention, if we accept the analogy I've sketched above, based on the *sub specie boni*-principle. There's something fishy with the attitude expressed by 'x is bad, but I intend it', while it is deplorably frequent that we find ourselves, as members of a team or some group, in a position to think 'x is bad, but we intend it'—this is especially obvious to citizens of direct democracies, or perhaps to teenagers who have tried in vain to convince their parents of what makes for a truly good holiday. The decision we've made is *ours*, and thus is an intention. A proponent of some minority position is not thereby excluded from the group that has the intention; whatever commitment is involved in the group attitude involves them as well. But any such member may be firmly convinced that the intended course of action is utterly bad. Tough as it may be to be in this situation, there is clearly nothing paradoxical about it. Again, there is no parallel between the case of individual and collective intentional action, as far as the features of the 'groundless knowledge' of what it is we're doing is concerned. Right as Laurence may be in observing that joint action involves reasoning from a shared point of view, it is obvious that joint intentions or shared beliefs do not commit us *in the same way* as individual intentions and beliefs do, and the difference at stake here seems to be a fundamental one, not one of degree.

(d) This brings us to the last feature of the 'deep structure' of groundless self-knowledge, the issue concerning authority. In the above sketch, a self-blind person was characterized as a person who refers to him- or herself only in the third person. Now it is obvious that even though we do have a first-personal way of referring to our groups—the collective use of 'we'—the third-personal approach to a group to which one belongs seems quite normal, such as when we say 'the department does x' rather than 'we do x', and that way of expressing group attitudes seems to reflect a difference between two ways of thinking of those attitudes. If we call illeists (from latin: *ille*) those individuals who refer to themselves only in a third-personal way, we should call illiists (from *illi*, the plural of *ille*) those who refer to their groups only in a third-personal way (cf. Schmid 2014b). Illeism does occur—American basketball players are famous for it, and parents seem to be prone to it, too ('Daddy's going to pick you up')—but it is rare, and if it were not just a manner of speaking, but a question of the kind of self-knowledge a person has, it would certainly be pathological. Illiism, however, seems to be frequent, and perhaps the standard case. Some people know that they are members, but they simply do not take a first-person-plural stance. They may go along with whatever group attitudes require them to do, but they do not 'identify' with the group perspective in the way that is expressed in the subject use of 'we'. It is not initially obvious at all that such attitudes on the part of the group members should make it impossible for agents to act jointly. Thus it seems that any such attitude is not necessary for joint action. Granted, there are cases of individual action that are similarly alienated—for example, when a person is coerced by another person, or acts on desires which she disowns, such as in the case of strong addiction. In such cases, the agent does not see herself as the 'author' of her action in the way expressed in the subject use of 'I'.[1] But this parallel highlights the basic difference between the singular and the plural cases—what is quite normal in joint action, if the above remarks are correct, is certainly an unusual condition in individual action.

Another question is if group identification, where it occurs, mirrors first-person-singular authority. Imagine a person who is closely identified with his group and does a lot of his reasoning in plural terms. He does not see his group membership in terms of an external circumstance, but as a part of his identity. It seems clear, however, that he does not thereby acquire any sort of special epistemic authority concerning what it is the group thinks, intends, or is doing. His position with regard to the group's intention is not privileged *in principle*, compared to an outside observer (for an extended version of the following argument, cf. Schmid 2014b). Even a person who conceives of his whole life in terms of his membership may misunderstand his group's attitudes, as any church member who has made the effort to find out what it exactly is that 'we' believe knows all too well.

If group identification does not constitute first-person plural authority, however, perhaps authorization does. List and Pettit (2011, 35 f.) argue that some

---

[1] I owe this example to Miranda Fricker.

groups, or corporate agents, have spokespeople that are endowed by the group with a special authority to express the corporation's attitude. Of such people we assume that they are in a privileged position to know what the corporation believes and intends. But there is serious reason to doubt that authorization by a group leads to first-person plural authority. There may be good reason to trust a spokesperson more than any other observer, but that does not imply a difference in the *kind* of knowledge a spokesperson can rationally be assumed to have, as opposed to an observer. Even the best-informed spokesperson for a business enterprise has to rely on information concerning the board's decision, and any journalist in the audience of his or her press conference may happen to have better information.

The only promising candidate for being in the kind of 'know' that is of truly first-person authoritative kind would be some sort of dictator, or perhaps an all-powerful CEO who also happens to own the company and dominates its board of trustees. In that case, it is really *up to him or her* what the corporate attitude is. He or she is calling all the shots concerning what it is the group is doing. In that case, the corporate intentional action really does reflect his or her rational stance, and the authority structure highlighted by Moran and others is fully in place. Such cases may be possible, but by the same token, any such example would not be a case in point; it would not be a case of *joint* intentional activity. Rather, it would be a case in which one agent's capacity for intentional action is extended; in the way in which it is intended by that authoritative individual, the action in question is not the action of a plurality of participants teaming up in acting together, but rather the individual intentional action of one individual that has increased powers because he or she can count on other people to do whatever she or he wants. Thus, instead of an analogy we seem to end up with a contradiction: wherever there is first-person plural authority, there cannot be *joint* intentional activity.

3. Plural Pre-Reflective Self-awareness

In the light of the above considerations, speaking of groundless self-knowledge of intentional joint action, or groundless group self-knowledge, does not just appear as 'laying it on a bit thick' (Laurence 2011, 293). It seems outright wrong, and whatever remains of the 'Anscombian approach to collective action' may seem like it cannot amount to much more than a faint analogy. I think, however, that we should resist this conclusion, and that it is now time to turn the tables. The previous section zoomed in on the dis-analogies between the singular and plural case with regard to the features we have identified as elements of groundless self-knowledge. But there is a positive side to explore, too. True as it may be that the knowledge of the participants in joint action cannot be of the exact same sort of knowledge of individual action that Anscombe labels 'groundless', it seems equally wrong to equate the knowledge in question with pure perceptual knowledge. And with a bit of good-will, one can even find some leads towards an account of genuine groundless group self-knowledge in the received literature. That it is not exactly *the same* should not come as a surprise. After all, the plural is not just another singular.

(a) First-person plural identity: the paper in which the label 'collective intentionality' was introduced in the debate ends with a section on 'Presuppositions' (Searle 1990, 414 f.). Searle here argues that it is not enough for people to engage in a joint intentional activity just to know one's own intention, and to know what it is other people intend, and to know that there is some structure of mutual awareness between the agents. Collective intentionality, he claims, involves a pre-intentional 'sense of "us"', an 'awareness of oneself and others as potential or actual cooperators', and he even uses terms such as 'sense of community' or 'communal awareness'. I have followed these leads elsewhere, and I have argued that the 'sense of "us"' in question here is plural pre-reflective self-awareness, that is, the participants' awareness of what it is they're doing together as their joint action, collectively. The 'us' is not of the sort that it figures in the content of the attitude in question. Rather, it is the *subjective aspect* of the consciousness in question, that is, the way in which whatever 'what it is like' there is to consciousness also involves a 'for somebody'. In the case of plural consciousness, that subject is not 'me' but 'us' (Schmid 2014a). Pre-reflective self-awareness is explained in contradistinction to self-knowledge, and the decisive difference is in the way each of the attitudes is 'of' oneself. Self-knowledge is 'of' the self in terms of the self being the object or 'theme' of one's attitude, such as in the case in which one thinks how a new hairstyle would suit one's features, while self-awareness is of the self in the way of the subjective aspect of consciousness. The distinction at stake here has been analyzed most clearly by Jean-Paul Sartre in his 1947 paper on '*conscience et connaissance de soi*'. Sartre's distinction fits Anscombe's distinction between the groundless and the perceptual kind of knowing oneself seamlessly.[2] The central idea that I take to be implied in Searle's remarks on the 'sense of "us"' is that Sartrean pre-intentional self-awareness does not only come in the singular. It sometimes comes in the plural, too. Sometimes it

---

[2] Moran has clearly missed a chance here—he draws heavily on Sartre, but he quotes only from his earlier *opus magnum*.

is our pre-intentional sense of our *individual self* that comes with whatever it is we're experiencing—if you're reading a book, the reading experience is transparent to you *as yours*. If we're engaged in joint action, however, the pre-intentional sense in question is likely to be of the plural sort: if we are walking together, the walking experience is transparent to us as *ours*, collectively. And true as it may be that no such 'sense of "us"' is infallible, it is immediate, nevertheless. It is that pre-reflective sense itself that answers the question who it is that the agent takes to be in and out, not some knowledge of other people.

(b) First-person plural perspective. In the previous section it was argued that Shoemaker's problem of self-knowledge does not carry over to the plural case. But that is only half of the story. There is such a thing as 'plural self-blindness', that is, the incapacity to see the attitudes one shares with others first-personally. Let us draw the parallel as closely as possible, and define as plurally self-blind a creature who has the conception of the various mental states a group may have, and can entertain the thought her group accepts this or that view as true, and has this or that intention, but who is unable to become aware of the truth of such a thought except in a third-person way (say, for example, by checking the minutes of the last group meeting). Other than what she knows in that way she cannot 'see' as a jointly accepted attitude. She has no other sense of what her group wants and accepts, that is, she has no pre-reflective 'sense of "us"', but only a sense of her individual self, together with observations of decisions and actions of her group, which she takes to be 'hers' only in the way of an illiist sketched above.

Contrary to the singular case, first-person-plural blindness does certainly not disqualify an agent as a reasoner. However, it seems that any such condition does disqualify him or her as a *joint reasoner* on two accounts. First, joint reasoning is a cooperative process of discussion over what it is we should accept as true, and what we should do. However, the question of what should be done cannot be approached cooperatively from the point of view of a person who is plurally self-blind, because that person can take only his or her own attitudes as a base of whatever reasoning process in which he or she is engaged. For a plurally self-blind person, other people may be valid sources of information, and the discussion with them may serve as an important means to test and improve his or her own attitudes, as well as to influence those other people's views. However, using others as sources of evidence, or as targets of one's influence, is not *partnership* in a cooperative reasoning process.

From the point of view of a plurally self-blind person, the only possible answer to the question what a group should do is the following: the group should do whatever *I* think is best (for the group), in the light of the reasons I have. In contrast, joint reasoning is predicated on the assumption that what we (together) should do is what *we* think is best in the light of the reasons *we* have. From a first-person plural perspective, the participants should articulate what they individually think is true or good, but it is only on the base of agreed-upon attitudes that the group has reasons to act, and this is what a plurally self-blind person cannot see. However benevolent towards others a creature with plural self-blindness may be, and however rational he or she may be as an individual, he or she simply cannot engage in joint reasoning—in that sense, the singular and the plural case seem to be parallel after all.

(c) First-person plural commitment. The claim that a plurally self-blind person cannot be a joint reasoner becomes clearer if we move on to the issue of commitment. Take again the person who is plurally self-blind, and has only third-personal knowledge of the attitudes of the group of which he is a member—for example, he registers the majority decisions that are made in his group, and he knows that the way in which group decisions are made is based on that process. In many cases he will find himself to have views of the Moorean kind: 'p is wrong, but we believe it'; 'p is bad, but we intend it'. For a person who is plurally self-blind, this is just another case of different and perhaps mutually incompatible attitudes held by different agents. Thereby, he fails to see the situation for what it really is: a *problem* that needs to be worked out in joint reasoning. It is true that divergence in opinion between individuals and groups always exist, even in healthy democracies; but it is equally true that a healthy democratic culture cannot survive for long if nobody cares in the slightest about whether or not what the group believes or intends and what he or she believes or intends are in line. Whoever thinks that what we believe or intend is wrong or bad should speak up and initiate a process of joint reasoning. Attitudes like 'p is bad, but we intend it' are not paradoxical. But among joint reasoners, such cases certainly mark a *tension*, or a deficiency of the relevant reasoning process. The situation is not unlike the case in which an individual reasoner makes up his or her mind, but finds some parts of his or her own mind remaining unconvinced of the stance he or she has taken. A plurally self-blind person fails to see a situation in which his or her view differs from the group's in the Moorean way as being of *that* kind of a problem. It is true that any real democratic process, especially in larger groups, demands of us that the joint reasoning be terminated at some point, and that the joint decision be accepted even by those who disagree. But joint deliberation would not even get off the ground if nobody perceived a contradiction between what he or she thinks and intends and what his or her group accepts and intends as more than just any divergence of opinion. For a plurally self-blind person, there is no way to see that what 'we' think and intend should ultimately be what *each of us* has reason to think and intend—without any further reason such as fear of nonconformity, loss of reputation, sanctions, or some such, but simply in virtue of the attitude in question's being *ours*, collectively.

A plurally self-blind person identifies with the group only in a third-personal way: 'There is the group (of which I know I'm a member). And here is where I stand'; and she fails to identify with the group in a first-personal way: 'The group is *us*.' One of the places at which this issue of first-personal identification with the group has been addressed in the received literature is in the last chapter of Philip Pettit's and Christian List's book on group agency. The authors argue here that group members need to think of their groups in we-terms. They need to collectively self-identify as a group agent, that is, to recognize that the group agent is 'us' (List and Pettit 2011, 186 ff.). They need to see the group attitudes as *their* (joint) commitments in a way that does not open what Pettit and List call the 'identification gap'. The identification gap is between what one believes the group to intend or believe, and what one takes to be one's own concerns: this is what that group intends to do—but why should *I* bother? Pettit and List see two ways of closing the identification gap. The first is based on a further thought. It requires of the individual members that they are disposed to do what the group attitudes require, based on the knowledge, 'well, *I'm* a group member after all'. According to Pettit and List, this is not enough. Identification cannot be left to any such extra 'cognitive achievement'. The identification has to be a 'by-product' of the group attitude itself—the 'by-product model' is the preferred way to close the identification gap. In that model, there has to be a feature in virtue of which it is a direct consequence of the attitudes of the group that the participants are disposed to go along. Put in other words, the group's attitude has to figure as an element in the individual's deliberative base in a way in which it is not dependent on the agent's thought that he is a member of that group.

Here is how Pettit and List imagine such a 'by-product model' identification with the group to work:

An analogy may help to explain the idea. Consider how pilots relate to cockpit instruments that give information on the plane's altitude, orientation, speed, and so on. Beginner pilots may consult the instruments and let the evidence count in their own reasoning process. If the horizon is out of sight, they may let panel information weigh against the information from their own senses; the latter may be misleading, since the senses cannot distinguish, for example, between gravity and acceleration... Rather than sustaining a relationship whereby the instruments are consulted, experienced pilots develop a different, direct connection with them. They let their own bodily cues go off-line, and hitch their intuitions and instincts directly on the instruments. They let the instruments guide them without the intrusion of thoughts about the evidence provided. Or at least they do this when the 'red lights' are off, and nothing indicates that things are amiss. Just as pilots can connect in this direct way to the instruments on the cockpit panel, so the members of a group may connect themselves directly to the attitudes of the group. They do not treat the group attitudes as mere indicators of what the group is to do, asking themselves explicitly whether they wish to identify with the group, and acting only if they have this wish. Rather, the individual attitudes are under the automatic guidance of the group, so that they can respond as spontaneously as pilots do when they take their cue from the panel before them. Or at least they may do this where there are no 'red lights' that suggest they should hesitate and take stock. (List and Pettit 2011, 192)

The basic idea of the problem of the identification gap, and of the 'by-product' model of closing it, is exactly right and important, but the way this is further analyzed in the above quote is certainly problematic, and here is why: to the degree to which an individual is accustomed to base his reasoning exclusively on another agent's beliefs and desires, he or she is not intentionally autonomous— his or her behavior does not instantiate his or her own action, but rather the action of the agent whose primary reasons rationalize that individual's behavior (cf. Schmid 2009). The way in which the participant's disposition to go along cannot be *that sort* of a by-product of the group's attitude, or else *only* the group, and not the participating individuals, would be agents. And even in List's and Pettit's account, this is not *really* what is meant, because List and Pettit emphasize that it is only in the light of the individual's knowledge that there are 'no red lights' on that they simply trust the group attitude. Therefore, what List and Pettit are suggesting here is not really a by-product model after all, but rather a version of the cognitive achievement model. The idea is that whenever people think it is okay for them to go along, they are disposed to act on the group attitudes. But this model uses the very element of which the by-product model was supposed to be free, namely, an individual attitude in light of which it makes sense for the individual to go along. So the dilemma is: either this account undermines intentional autonomy, which makes it implausible, or it isn't really a by-product model of collective identification.

The question is: is there a way of conceiving of collective identification without further cognitive achievements that is compatible with individual intentional autonomy? I submit that there is. The way collective self-identification is a by-product of an attitude is in virtue of the individuals' awareness of an attitude as *theirs*, collectively. The plural self-consciousness involved in this is not a *further* attitude that is added to the awareness of whatever it is that is in the focus of the consciousness in question. If I'm aware of our walking together *in the right way*, that is, first-person *plurally*, the awareness of that activity as being *ours* is part and parcel of the awareness of the walking that's going on. At the same time, it clearly does not undermine the status of my contributive action as *mine* if I do my part based on that first-person plural knowledge of what it is we're doing. I'm not acting on the ultimate base of an attitude that is another agent's, and not my own. I'm acting on attitudes that I *share*, that is, attitudes that *I have*, only that the way in which I have them is *jointly with others*. In that way, plural self-awareness of an attitude as *ours* closes the identification gap without further cognitive achievement, without thereby disqualifying the participating individuals as proper agents.

(d) First-person plural authority. The last point of comparison is the problem of first-person authority. This may seem to be the most serious obstacle to the idea of groundless group self-knowledge. As argued above, no single individual in a group can rationally aspire to being in a position to express her group's attitudes in a way that comes with true first-person authority. All a member can express with authority is *that she thinks* that the group wants, or believes, or intends to do—because whatever it is she pre-reflectively takes to be the attitudes she has jointly, with others, is not only up to her, but up to others as well. If a single individual were in a position to 'make up' her group's mind, as it were, and thus speak that mind with first-personal authority, that group would not be the agent of a joint action, but simply the extension of her individual agency. One way around this would be to claim that while no individual can express (and thus *know*) her group's attitudes, perhaps the *group itself* can. Ideas such as Raimo Tuomela's 'chorus sense of "we"' come to mind here, the case in which the group's attitude is expressed *jointly by the group members*, instead of just a single individual. Among the doubts I have concerning the view that some such collective speech act expresses a group attitude with first-personal authority is that it is not clear by the mere fact of the collective speech act that the participant individual speakers thereby have the same 'we' in mind (for a more careful argument, cf. Schmid 2014b). And that, of course, carries over to the kind of knowledge at work in these cases. No fully authoritative knowledge of what it is we're doing together is available to us, however aligned our views thereof may be. But the fact that there is no such thing as full first-person plural authority certainly does not prove that any knowledge we have of what it is we're doing together is third-personal. It does not even show that knowing what we're doing together comes with *no* first-person plural authority at all. I submit that a weaker form of first-person plural authority is widely accepted even in our everyday practice. We certainly do not grant the members of another group a privileged position as to the question of what it is their group is doing simply because we assume they are in a better position to observe what is going on among them. Rather, we grant them that privileged position because the attitudes of their group are what they, jointly, make them to be, and are, thus, they are co-authors of their attitudes in quite a literal sense. The disanalogy shrinks to the insight that the 'making' of a group attitude does not come with the sort of centeredness that is typical of rational individual minds.

One reason why the first-person singular viewpoint is granted special authority is the alleged impossibility of misidentification. My predication of what it is I think or intend may be wrong (such as in the case in which I mistake my temporary infatuation for true love), but the identification that it is *my* attitude cannot be mistaken. And it seems that there is no analogy for this in the plural case: I may easily be mistaken in assuming that an attitude is really *ours*. Here, my strategy is that of companionship in guilt: just as a person may be mistaken in taking an action or intention to be joint, when it is really just his individual own, he may be mistaken in taking an action or intention to be his individual own, when it is really joint. Take the case of you and me deciding to write a paper together. The process is a prolonged one, and we rarely talk about the progress we're making in developing the argument of our respective parts. As I am a bit egocentric, however, I somehow come to see the project as my individual own after a while, and that is how I see what I am doing: 'I am writing this paper.' Yet I'm simply mistaken in the identification of the subject of what I am doing; I am misattributing the intentional action, or even just the intention, to *me*, when it is really *ours*, as I come to realize on the occasion of our next conversation. In that way, I was mistaken in the subject of the action in question. Mis-identification does happen in the plural, yet thinking about joint action makes us realize a way in which it may happen in the singular, too. If first-person authority is an impossibility of mis-identification, it exists neither in the plural nor in the singular.

It is time now to come to a conclusion. Where does all of this leave us? I have argued that the idea of groundless self-knowledge of what it is one is doing can be spelled out in terms of first-personal identity, perspective, commitment, and authority. I have argued that, in spite of the obvious differences between the singular and the plural cases, the idea that the participants have groundless knowledge of what it is they are doing together makes good sense. There is first-person plural identity, perspective, commitment, and authority at work in joint action, and thus groundless group self-knowledge. Yet there are remarkable differences between the singular and the plural form of groundless self-knowledge. This should not be surprising. After all, the 'we' is no collective 'I'.

PART II
Ethics

4

The Social Epistemology of Morality
Learning from the Forgotten History of the Abolition of Slavery

Elizabeth Anderson

1. Learning from History

What country was the first in the modern world to permanently recognize an absolute, universal human right against slavery? France, with its 1789 Declaration of the Rights of Man and Citizen? No. It did not permanently abolish slavery until 1848. Great Britain, with its Slavery Abolition Act of 1833, ending slavery in the British Empire? No. Several states in Latin America, including Mexico, Gran Columbia, Chile, Uruguay, Bolivia, and the Federal Republic of Central America, preceded it. But they were not the first. The first country was Haiti, the site of the only successful slave revolt in world history. The self-emancipated people of what was then the French colony of Saint-Domingue had presented their freedom as a *fait accompli* to the French National Convention in 1794, under conditions that induced France to legally endorse their action. Napoleon reversed the Convention's abolition decree in 1802, however, and undertook a genocidal campaign to reimpose slavery in the French Caribbean colonies. The Haitians defeated Napoleon and declared independence on 1 January 1804, along with the unconditional abolition of slavery.

Did you answer this question correctly? I expect that most readers of this chapter, if they are citizens of a Western country and not historians, would not have done so. The dominant narratives Western countries tell about themselves is that they took the lead in advancing human rights throughout the world. The West has achieved enough self-awareness to recognize its own capacity for mass human-rights violations in slavery, imperialism, the Holocaust, and other crimes against humanity—although it has forgotten many of its crimes. In the dominant Western historical narratives, however, the West has forever been an autodidact, arriving at the true principles of morality through its own self-sufficient reasoning, figuring out for itself when it has failed to apply them, self-correcting its course, and taking the lead in teaching these principles to the rest of the benighted world. It does not imagine that it had to learn fundamental moral truths from those whom it victimized, particularly not from people of African descent.

That the truth is quite otherwise carries important lessons for moral epistemology. In this chapter I aim to explore how social groups learn moral lessons from history, particularly from their own histories. How do historical processes of contention over moral principles lead groups to change their moral convictions? My interest is normative: I am interested in changes in collective belief that count as moral improvements, as acquisitions of genuine moral knowledge. I aim to draw some lessons about how the social organization of moral inquiry—of contention over moral claims—affects the prospects that a group will be able to improve its moral beliefs.

By "morality" I refer to the domain of ethics that concerns what we owe to each other. This is a matter of interpersonal morality, of duties and obligations to others, that others have claims upon us to discharge. Such principles are supposed to govern not just moral thinking, but also social relationships. The question of the correct principles of moral rightness comes down to the question of which classes of interpersonal claims have authority over the conduct of those to whom they are addressed. This sense of morality includes norms that may be enforced by law or informal social sanctions.

A social group can be said to have learned a moral principle and hence to know it, only if the principle is operative within the group. This does not mean that every member personally believes it, much less that everyone obeys it. Rather, a group shares a conviction about a principle only if that principle shapes discourse within the group in particular ways: it is taken for granted within the group as a premise for further argument, not needing independent justification; its truth is treated as a settled matter; disputing it is regarded as, if not crazy or beyond the pale, then requiring a heavy burden of proof; disputants are liable to censure or even social exclusion for calling such convictions into question (Gilbert 1987). Because moral principles regulate interpersonal relationships, to count as a shared conviction of the group it must also shape conduct: members are free to make claims under the principle and generally do so when they are victimized by violations of it; other members acknowledge the legitimacy of such claims; the principle is widely, if not completely, obeyed by group members; the group punishes disobedience; members take steps to transmit the principle to future generations.

History is a resource for our epistemological investigation not merely because processes of group belief change are recorded in history, but because groups use their histories as a basis for drawing moral conclusions. They draw lessons for current practice from their interpretations of the past. On a pragmatist account of how this works, people learn about morality from their experiences in living in accordance with their moral convictions. We advance moral principles to solve recurring problems in our social lives. When circumstances change, those principles may no longer solve these problems, or new problems may arise for which they are unequipped. This may trigger fresh moral inquiry, a search for new principles (Dewey 1922). We may also envision better possibilities, new moral ideals that appeal to us more than the old ones. These, too, must be tested in experience to see whether their reality lives up to our dreams. Moral principles are tested in practice; experience with the results may either reinforce the principle or undermine it. This can take place at both individual and group levels.

Suppose a group, as a result of some historical process of contention over a moral principle, changes its moral convictions. How can the group tell whether this change amounts to moral learning—the acquisition of moral knowledge—or a moral mistake? It can, as already suggested, see if it finds social life governed under the new principles more satisfactory than life governed under the old— whether the new principle resolves longstanding interpersonal or inter-group conflicts better than the old, or replaces intractable conflicts with more tractable and less dangerous ones, or produces new benefits. It can take a long time to learn how to live under new principles, however, and the transition costs may in the short run obscure the long-run benefits of a new moral regime.

A second way a group may be able to tell that it has made moral progress is epistemological. Suppose we have an idea of the sorts of social arrangements that are liable to produce moral error, confusion, bias, or blindness. And suppose the historical processes that led to the group's change of moral conviction enhanced the group's epistemic powers—say, by blocking or overcoming certain sources of bias, or exposing it to new sources of morally relevant information—and that this enhancement helps explain the group's change of moral conviction. Then we may have good reason to think that the change of group belief amounts to genuine moral learning (Taylor 1993).

Either path underwrites a naturalized social epistemology of moral learning. We are to investigate how groups go about changing their moral convictions, looking for characteristic social sources of error and bias, and how groups may overcome, or fail to overcome, their epistemic defects. We are to see how open groups are to recognizing the problems that their moral convictions generate, and to revising them in ways that effectively address those problems. We turn to history to learn how groups have learned.

To understand how this works, we need to distinguish the perspective of the group from our outsider's perspective (even if we are historically connected to historical group in question, and so may be inheriting some of their views and biases). A group may take itself to have learned certain lessons from its history of putting certain principles into practice, while we may judge that the group drew the wrong conclusions from its history.

We also need to take special care to tell the epistemic story accurately—to be meticulous about the social processes by which a group changed its convictions. As the opening of this chapter indicates, groups may be tempted to assume they have learned moral lessons all by themselves. Yet social groups rarely acquire the conviction that they are committing wrongs against others from their own epistemic resources alone. Being called to account by the victims of their injustice is critical to the development of moral consciousness in social groups. I shall argue in this chapter that establishing the social conditions of accountability is critical not only for ensuring that agents comply with known moral requirements, but for their ability to learn what those requirements are. Sound moral inquiry is not only essentially social; it demands the participation of the affected parties, of those making claims on others' conduct, as well as those to whom such claims are addressed. We cannot hope to get our moral thinking straight unless we include the affected parties in our moral inquiry, and include them on terms of equality. The social epistemology of moral inquiry is, in a sense I shall define at the end of this chapter, essentially democratic.

Rather than attempting to establish this claim a priori, I propose to show how moral errors tend to arise when moral inquiry takes an authoritarian form. By "authoritarian" I do not refer to the explicit content of the moral principles arrived at, but rather to the social relations within which moral inquiry proceeds. Moral inquiry is authoritarian when: (1) it is conducted by people who occupy privileged positions in a social hierarchy; (2) the moral principles being investigated are those that are supposed to govern relations between the privileged and those who occupy subordinate positions in the social hierarchy; and (3) those in subordinate positions are (a) excluded from participating in the inquiry, or (b) their contributions—their claims—are accepted as requiring some kind of response, but where the response of the privileged fails to reflect adequate uptake of subordinates' perspectives, but rather uses their social power to impose their perspective on the subordinates.

I will explore the problems of authoritarian moral inquiry as it arose in Euro-American moral inquiry concerning slavery and the aftermath of emancipation. Here is a case about which it is evident that social groups have undergone historical processes of moral learning. Three centuries ago, Europeans believed that slavery was just. Starting about 250 years ago, this belief began to be actively contested on an international scale, not only in philosophical tracts or in the personal convictions of individuals, but in social and political movements attempting to change the moral beliefs of whole nations. Over the course of the nineteenth century, the belief that slavery is wrong became accepted by all of the countries of Europe and its colonies or former colonies in the Americas, and nearly all the rest of the world as well. It is impossible to summarize such a protracted process in a single essay. I therefore select two episodes in the history of contention over Euro-American slavery. The first took place in revolutionary France, when the first proposals to abolish slavery were being advanced. The second took place in the post-emancipation context, when people were testing abolition by its results. We shall examine them with the aim of discerning sources of bias or moral blindness in social groups, and how they were or were not overcome.

2. Episode 1: Bias in Enlightenment Proposals for Gradual Emancipation

European and American thinkers of the seventeenth and eighteenth centuries sometimes advanced arguments on the basis of purportedly universal claims about human rights and human nature. Yet some, such as Jefferson and Locke, either owned slaves or invested in the slave trade. Others, such as Rousseau, denounced slavery in the abstract but never specifically condemned the European enslavement of Africans. What accounts for these contradictions and silences?

It might be supposed that these thinkers were simply going along with group convictions of Euro-American states and colonies, embodied in the laws, that denied that Africans or slaves were entitled to claim any rights at all. Blacks were, in the notorious words of *Scott v. Sandford* (60 U.S. 393, 407) in 1857, "so far inferior that they had no rights which the white man was bound to respect." Yet the laws of slavery did recognize slaves as bearers of legal rights, even though slaveholders violated those rights in practice. Louis XIV's 1685 edict regarding the treatment of France's slaves, the *Code Noir*, recognized that slaves had several rights against their masters, including the right to practise the Catholic faith, to rest on the Sabbath, and to be provided food and care when too old or infirm to work. They also had rights to family integrity: against being forced to marry anyone without their consent, and against family members being sold separately by their master. Masters could free their slaves for any reason. Once freed, former slaves were legally entitled to the same rights as the freeborn. France thereby acknowledged that there was nothing inherent in the nature of those who had been enslaved that made them ineligible for equal rights. American slaves, too, had certain legal rights, notwithstanding *Dred Scott*. In the US context, the practice of holding slaves legally responsible for their crimes accorded them the procedural rights of criminal defendants under the common law, including the right to a trial and to bear witness in their own defense. Some courts in the South even recognized slaves' right to use force against their masters in self-defense (Oakes 1990, ch. 4).

Nor did the advocates of universal natural or human rights make a racist exception to justify slavery on moral grounds. Jefferson, although a slave-owner and a racist, knew that slavery was unjust (Jefferson 1905, Query XVIII). Locke, although he invested in the slave trade, argued that legitimate slavery was limited to the combatants in an unjust war, and denied that their countrymen or descendants could be justly enslaved (Locke 1980, sec. 179). That is utterly incompatible with the practice of chattel slavery in the colonies. We encounter nothing so simple among the leading Enlightenment thinkers as a syllogism with a true major premise and a false, racist minor premise to justify slavery. The corruptions of moral thinking involved in Enlightenment thought concerning slavery are typically more subtle, and tied to an authoritarian moral epistemology.

Early proposals to abolish slavery within France offer a useful illustration of the problem. In 1790, Baron de Viefville des Essars (1790) submitted an emancipation proposal to the National Assembly. Despite his condemnation of slavery as a violation of the slave's inalienable rights, he did not insist on immediate abolition, but planned for a gradual emancipation process extending over sixteen years. The Assembly ignored his proposal. The Abbé Grégoire, too, argued for gradual emancipation (James 1963, 141). One might see the point of gradualism if this were merely a concession to feasibility. France was economically dependent on slave labor. Saint-Domingue, by far the richest colony in the New World, produced 40 per cent of the sugar and 60 per cent of the coffee consumed in Europe (Dubois and Garrigus 2006, 8). About one out of twenty-five people in France directly depended on trade with its colonies. Many of the richest men of France owed their wealth to their ownership of plantations, participation in the slave trade, or marketing the products of slave labor. Among the members of the National Assembly, 15 per cent owned property in the colonies (Dubois 2004, 21). Any emancipation program could therefore expect to meet overwhelming resistance if some provision were not made to compensate the slave-owners for their massive investment in slaves, and to develop the institutions needed for a transition from slave to free cultivation of cash crops. On their own natural-rights principles, the abolitionists would have to admit that the temporary continuation of slavery for these reasons was unjust. But the concession could have been excused as the only feasible path to emancipation, given the powers lined up in favor of slavery.

Nicolas Condorcet's work shows how matters were more complex than this. Condorcet was both a feminist and an opponent of slavery. He belonged to the abolitionist Société des Amis des Noirs. Yet there is a striking contrast between his feminist and his abolitionist arguments. His argument for equal rights for women displays perfect syllogistic reasoning:

[T]he rights of men result simply from the fact that they are rational, sentient beings, susceptible of acquiring ideas of morality, and of reasoning concerning those ideas. Women having, then, the same qualities, have necessarily the same rights.
(Condorcet [1790] 1912)

The argument is straightforward. Condorcet goes on to refute objections that women are incapable of exercising the rights of citizens. There is no suggestion of backtracking, reservation, or exception in his argument, nor any reason to delay implementation of his conclusions.

Compare this to Condorcet's argument in *Réflexions sur l'esclavage des negres* (1781). It starts off well enough, insisting that slaves have been unjustly deprived of the right to dispose of their own persons. Justice therefore requires the abolition of slavery. Condorcet never suggests that slaves or blacks lack any human rights. He dismisses numerous excuses for slavery. Free labor can produce sugar just as well as slave labor. Even if slavery were the only way to produce sugar in the colonies, gluttony could never justify theft of another's labor. Slavery is such an outrageous crime that slave-owners are not entitled to compensation for freeing the slaves, any more than thieves are entitled to compensation when their stolen goods are confiscated.

Condorcet's reasoning would seem to imply that, just as in the case of women, slaves should enjoy immediate emancipation. Yet Condorcet proposed an extraordinarily protracted abolition process. While the slave trade would be abolished immediately, emancipation of already existing slaves would be phased in by age. Infants born to slavery would be required to serve their masters until they reached the age of 35. Children under 15 would be bound to their masters until the age of 40. Slaves over 15 would not have their freedom until the age of 50. In part, delay was needed to

give time both to the colonists to change their farming methods gradually and secure the means necessary to cultivate their lands by employing whites or freed blacks, and to the government time to reform the laws and policing system of the colonies. (Sala-Molins 2006, 14)

More ominously, he claimed that protracted emancipation would be needed to prevent the masters from violently attacking their former slaves. We must recognize:

> the danger to public order posed by the fury of masters wounded in their pride and avarice—for a man who has been used to seeing himself surrounded by slaves will not now be easily consoled by being surrounded by mere social inferiors. It is considerations such as these that can allow the legislator to defer, without committing a crime, the repeal of any law that deprives another man of his rights. (Sala-Molins 2006, 20)

These were not Condorcet's main reasons for delay, however. More importantly, he argued that slaves, having been unjustly deprived of their natural rights, were not ready for freedom. If granted the same rights as whites, they would form mobs and exact revenge against them. They would cast off civilization, fly off into the mountains, and live as vagabonds. The plantation economy would collapse, and the colonies would descend into crime and disorder. Since legislators are obligated to craft laws that they can reasonably expect will protect everyone's rights to the greatest degree possible, the slaves therefore cannot be allowed full freedom until they have learned to exercise responsibility. In the meantime they must be subject to "severe discipline".

Condorcet insisted that the slaves were not inherently incapable of governing themselves. Their incapacity was due to the fact that life under the master–slave relationship had corrupted them. They are analogous to:

> men who have been deprived of some of their faculties through misfortune or illness, and who cannot be allowed the full exercise of their rights lest they harm others or themselves.
> If . . . the slightest certainty exists that a man is unfit to exercise his rights, and that if he is allowed such exercise of them, he will constitute a danger both to others and to himself, then society is entitled to regard that person as having lost his rights or as never having had them. (Sala-Molins 2006, 18)

Condorcet's reasoning on this point is strained. Women, too, had been deprived of their rights under patriarchy, but Condorcet never inferred that they should continue to suffer disenfranchisement and subjection to their husbands because patriarchy had robbed them of the capacity for freedom. Nor can his reasoning justify the protracted enslavement of infants and children, who would be too young to have experienced the corruption in question. Condorcet was also aware of the fact that the colonies contained numerous free people of color, many of whom had been freed by their masters. A cursory examination of their conduct would have revealed no incapacity for freedom on account of their prior servitude. In fact, the free people of color of Saint-Domingue played critical roles in its economy, as independent farmers, merchants, and planters, and also supported the government in the militia and police force (Dubois and Garrigus 2006, 15). Even if we imagine that freed people would need some time to adjust to emancipation, it is hard to believe that further experience of slavery could prepare them for freedom. Condorcet's own premises contradicted that possibility: if enslavement was what had corrupted them, how could continued enslavement teach them to be free? How else can one learn to handle freedom responsibly, other than by freely making choices for oneself and learning from the consequences?

Even more astonishing was Condorcet's plan for teaching slaves how to handle their freedom responsibly. This job was to be assigned to their masters! Although the masters were guilty of the most monstrous injustices—violent conquest, tyranny, robbery, rape, torture, sadistic murder—they were now to be trusted with educating their slaves for freedom. Although their anticipated fury at being deprived of absolute dominion over others was one of Condorcet's justifications for allowing them to continue their mastery for decades, they were now to be charged with educating their slaves for the enjoyment of equal rights. Although the cause of the slaves' moral degradation and consequent incapacity for freedom was the fact that they supposedly followed their masters' example, knowing nothing better than their masters' uncontrolled greed, cruelty, sloth, and lust for power, now the masters were expected to teach them virtue:

> Considering the happiness of the slaves as his supreme duty, and the loss of their liberty and rights as an evil it behooves him to correct, he rushes to his plantations to shed his tyrannical ways, to don the authority of the just and humane sovereign to commit himself to making human beings out of his slaves. He trains them to become industrious workers and intelligent farmers . . . The vices of the slaves would disappear with those of the master . . . Honesty, love of virtue, maternal and filial love—all these tender emotions, enriching the life of these unfortunate people—become the fruit of his labor.   (Sala-Molins 2006, 24)

Louis Sala-Molins (2006, ch. 1), acidly reflecting on the passages quoted here, condemns Condorcet as a racist. My concern in this chapter is not to pass moral judgment on Condorcet's character. It is to query Condorcet's social epistemology of moral knowledge. Condorcet explicitly denied that blacks were unfit by nature for freedom and human rights. Yet implicitly he could not imagine former slaves teaching themselves how to handle freedom responsibly. He imagined that their white masters, notwithstanding their extreme viciousness and wrongdoing on an immense scale, would be needed to teach their slaves how to be moral. Nor did it cross his mind that the slaves and free people of color might teach their masters, or the French people more generally, something about moral right and wrong.

This is what authoritarian social epistemology looks like in practice. Despite the fact that Condorcet grasped genuine principles of moral rightness, his defective social epistemology, which assumed a one-directional line of moral instruction from Europeans to black slaves, led him to absurd contradictions. It was, of course, an epistemic injustice (Fricker 2007) to the slaves to exclude them from participation in moral inquiry into how slavery should be dismantled. This injustice corrupted Condorcet's own moral thought, as it did the moral thinking of European societies more generally.

Why did Condorcet think so much more consistently about feminism than about abolition? It is plausible that the fact that he was "in constant intellectual communion with his wife" (Schapiro 1963, 189), had something to do with his relative moral clarity (Nall 2008). His wife, Sophie De Grouchy, was the author of *Letters on Sympathy*, translator of Adam Smith's *Theory of Moral Sentiments*, and hostess to a prominent salon in Paris. Inclusion of the object of his feminist concern as a co-inquirer likely enabled him to think straightforwardly about women's emancipation. By contrast, Condorcet was isolated from the slaves in the colonies. The membership of the Société des Amis des Noirs, which endorsed only gradual emancipation, was highly elitist and segregated.

3. Overcoming Bias: Slave Participation in Enlightenment Contention over Slavery

This argument, that isolation of elites from engagement with the perspectives of subordinated groups corrupts their thinking, may be extended to dominant social groups in general (Anderson 2007; 2010, ch. 5). We should recognize, however, that the slaves were not wholly excluded from participating in Enlightenment contention over human rights, slavery, and emancipation (Dubois 2006). They were in a position to force their European and American masters to sit up and listen to their claims. Historians of slavery have led the way in reconceptualizing what subaltern participation in such contention amounted to. In some cases, letters, petitions, and other documents articulating slaves' complaints are preserved in the historical record. Issues of moral right can also be contested in action: slaves made their moral claims known through their patterns of resistance to domination. Successful resistance forced the masters to recognize that their claims to legitimate power were rejected, the imagined happiness of their slaves an illusion, their projects for securing social order tenuous at best. It forced them to diagnose what was going wrong with their practices, so that they could be altered in ways that would address the problem. Sometimes, it forced them to find fault in themselves, and not only in their slaves.

In the French Caribbean, resistance often took the form of marronage—escape from the plantations to the freedom of the forested hills, where slaves were difficult to track down and could eke out a living (Dubois 2006, 1–2). Their freedom of movement across different plantations enabled maroons to serve as conduits of communication among the slaves, and as potential instigators of revolt. The official French response was not simply to escalate violence and repression. Marronage forced the *métropole* to consider that any human being would try to escape cruelty. The *Code Noir* already recognized that masters needed to be restrained from torture and other gross abuses, lest the slave system unravel. In the 1780s, the Colonial Ministry responded to slave unrest by granting new rights to slaves, including the right not to work on Saturday afternoons and the right to issue formal complaints against their masters' abuses, potentially leading to punishment of the masters. The masters rejected these restraints on their power, as they had ridiculed the earlier provisions of the *Code Noir* (Dubois 2004, 30–1). But in Saint-Domingue the slaves were to have the last word, when they revolted in 1791.

In the American South, too, slaves exploited the tension between masters' will to total domination and the state's insistence that slavery be regulated by law. If carried out frequently enough, every mode of resistance—leaving the plantation without a pass, learning to read and reading seditious documents, breaking tools, defending themselves against their masters' violence—forced a legal response. In a legal system constituted by common-law rights to due process, slaves accused of crimes took advantage of those rights. Bit by tiny bit, slaves used the legal system to win new rights—even the right to defend themselves against their masters' excessive violence. "Every major court decision elaborating a slave's rights was instigated by some act of resistance on the part of the slave" (Oakes 1990, 165).

We glimpse here the outlines of an alternative social epistemology of moral inquiry. As in standard philosophical models, it is dialogic in form, consisting of claims and counter-claims, made on the background supposition that progress can be made through an examination of the merits and weaknesses each side's claims has in relation to the other. Unlike standard philosophical models, however, the dialogue is not merely imagined in a single person's head, or pursued by participants who are detached from the claims being made. Rather, the critical claims and counter-claims arise from interactions of the affected parties—those who are actually making moral demands on one another, and insisting that the others offer them a serious normative response. This alternative social epistemology is naturalistic. We are interested in how moral discovery actually takes place, and the conditions under which it happens. After the fact, it may be possible to rationally reconstruct a straightforward moral argument that encapsulates certain moral lessons. Before a group consolidates a consensus around such lessons, however, its path to discovery may be fraught with obstacles to moral understanding. It may take conflict among the claimants, even violent conflict, to clear away those obstacles.

In both the United States and Haiti, violent conflict was necessary to generate moral clarity about the urgency of abolition. In both cases, two types of action by slaves helped clear the moral sensibilities of those in power—self-emancipation, and serving their country in war. Service in war served two clarifying functions. First, it triggered gratitude among the white leadership of France and the Union, who saw that the survival of their social order depended on the voluntary actions of subordinates. Gratitude tempers the vanity of superiors and opens their ears to legitimate claims. Second, the slaves' demonstration of valor and skill in combat defeated the assumptions of incompetence and inferiority that rationalized slavery. The Confederate leadership recognized the moral force of such a demonstration. This is why it rejected black soldiers until manpower shortages towards the end of the war led to a desperate (and futile) attempt to recruit them. As Georgia Governor Joseph Brown (1865) put the point: "Whenever we establish the fact that they are a military race, we destroy our whole theory that they are unfit to be free." Howell Cobb (1865), one of the founders of the Confederacy, concurred: "If slaves will make good soldiers our whole theory of slavery is wrong."

It is well enough known that the slaves of Saint-Domingue freed themselves through revolution. In white Americans' historical memory, reproducing the autodidactic myth of the powerful, American slaves were freed by Lincoln's Emancipation Proclamation of 1 January 1863. Yet the Emancipation Proclamation would have had little effect had not the slaves emancipated themselves by escaping behind Union lines.

How does *de facto* emancipation promote moral clarity? When one group lives at the mercy of a dominant group, the dominant group has no need to resort to persuasion or bargaining to get what it wants. It can simply impose its will, giving little thought to moral claims issuing from the subordinate group. By contrast, when it confronts masterless people—those who not only have no *de facto* masters but cannot be subdued by force—it must resort to other strategies that appeal to their interests (Herzog 1989, ch. 2). This entails some recognition of the other as bearing legitimate claims. Such recognition amounts to a moral advance.

In 1794 the National Convention received the racially diverse representatives of Saint-Domingue, who brought news of emancipation, with joy. They ratified emancipation and extended it to the rest of the colonies by acclamation. Barely four years after France had rejected even gradual emancipation, it embraced immediate emancipation. More than one representative observed that doing so was required by the revolution's own principles. Danton noted, "until now our decrees of liberty have been selfish ... But today ... we are proclaiming universal liberty" (Dubois and Garrigus 2006, 131). Events can make it easier to immediately draw the logical conclusion from a practical syllogism.

4. Episode 2: Testing Theories of Moral Right in Practice. The Case of Emancipation

In contrast with the dominant methods of contemporary analytic moral philosophy, which test moral principles only in thought experiments, those engaged in contention over slavery believed that the case for emancipation had to be tested in practice. Because the fundamental justification for the slave system rested on the assumption that slaves were unfit for freedom, the test examined how the freed people exercised their freedom. Condorcet's plan for a protracted emancipation process had been based on the assumption that slaves would need to learn how to use freedom responsibly before being emancipated. What lessons were the freed people supposed to learn from their former masters? The answer those in power gave to this question—even those who, like Condorcet, thought slavery was unjust—goes to the heart of what they thought freedom was supposed to amount to for freed people.

The answer may be found in a pivotal moment in the history of emancipation, when Étienne Polverel promulgated labor policies for the abandoned plantations confiscated by France in the midst of the slave revolt in Saint-Domingue. Polverel and Léger-Félicité Sonthonax were Civil Commissioners sent by the National Convention to re-establish French control over Saint-Domingue and enforce a decree requiring equal treatment of free men of color and whites. Polverel governed the west and south of the colony, Sonthonax the north. In the course of the revolution both discovered that, to secure the colony for France, they had to win the slaves to their side by proclaiming emancipation. The freed people, however, had a different understanding of what freedom meant than the Civil Commissioners. We can infer their demands from the labor regulations Polverel issued for the plantations under state control in 1794 (Dubois and Garrigus 2006, 139–42). Polverel initially set the compensation for the freed people at one-third of the net revenue of the plantation, with two-thirds going to the owner. However, he explained, this share was based on the assumption that they would take only one day of rest. He assured them that, as free workers, they could choose not to work on Saturday as well. But they would have to bear the entire cost of taking their leisure. Each extra day of leisure reduced profits by one-sixth. Hence if they took Saturday off they would be paid only one-sixth of the profits instead of one-third—a 50 per cent pay cut.

The freedom of the landowner consists in the ability to have his land worked as he wishes, by whomever he wishes, and in the way that he wishes. He would start by evicting the entire lazy work crew from the plantation and hiring day laborers to work his land. He would no longer have to provide shelter or a provision ground to the field hands.... Africans, now you have been educated. Let us see if you will still choose to rest on Saturday.... (Dubois and Garrigus 2006, 141)

Polverel also rejected the demand of women workers for equal wages, and denied that the freed workers had any rights to the gardens that they had enjoyed as a customary right under slavery. This had been a convenience to the slave-owners, who did not have to purchase provisions for their slaves if the slaves grew their own food. In a wage-labor system, however, workers had to provide for themselves. While Polverel conceded limited gardens to the workers as a continuation of custom, he rejected their demand for larger plots. Denying the freed people independent access to the land was necessary to ensure that their only option for survival was to continue their work on the plantations, raising cash crops to the enrichment of landowners.

The leaders of all post-emancipation societies shared Polverel's view of the freedom to which the freed people were entitled. Sonthonax, in proclaiming emancipation in the north, instituted wages and banned the whip, but bound field hands to one-year contracts, limited their freedom to change plantations, and permitted punishments for violations of work discipline, including stocks and fines up to the worker's entire salary (Dubois and Garrigus 2006, 121–5). Toussaint Louverture's 1801 constitution for Saint-Domingue bound workers to the plantations of their former owners (Dubois and Garrigus 2006, 169). The British, upon emancipating their slaves in 1833, instituted six-year "apprenticeships" that required the freed people to continue working on the plantations under the rigorous discipline of their owners (Foner 2007, Kindle loc. 397). Planters in the American South also attempted to continue gang-style plantation labor for wages, and enacted the notorious Black Codes to enforce this system during the first phase of Reconstruction (Foner 2007, Kindle loc. 970–1004). In all these cases, the judgment about freed people's entitlements was driven by authoritarian moral inquiry. Polverel at least recognized that the freed people's moral claims needed to be addressed. However, he used his power to impose his preferred solution rather than giving their claims a serious hearing.

The freed people had a dramatically different conception of the freedom to which they were entitled. Across the post-emancipation societies, former slaves identified freedom with self-government, being one's own boss, not having to take orders from an overseer. They wanted not only self-ownership and the right to the fruits of their labor, but to decide for themselves how hard they would work, for how long, at what tasks, and under what conditions. No free person would willingly work under the brutal conditions, harsh discipline, and gruelling intensity required for generating maximum profits in the plantation system. That is why plantation owners consistently sought to raise cash crops by exploiting various forms of unfree labor—slavery, indentured servitude, debt peonage, serfdom. Given independent access to land, either through ownership, renting, or sharecropping, freed people virtually everywhere chose to reduce the hours and intensity of their labor, shift from cash crops to subsistence farming, and release women and children from intensive fieldwork so they could devote more time to family life. The struggle for freedom in post-emancipation societies thus became a struggle with landowners over access to land (Foner 2007). This had different outcomes in different societies, depending on the relative political power of landowners and freed people and the availability of open land. Across most of the American South, the outcome was a sharecropping system. In Haiti most freed people succeeded in turning themselves into an independent peasant class. Large numbers in Jamaica and Cuba also managed to establish themselves as self-governing farmers.

Were these outcomes a vindication of freed people's capacity for responsible self-government? The dominant conclusion of contemporary whites in post-emancipation societies was that the former slaves had failed the test of freedom. Even abolitionists regarded the decline of sugar production in Jamaica as a "serious embarrassment" (Foner 2007, Kindle loc. 629). They disparaged the Haitian peasantry for failing to manifest the mental progress supposedly entailed by embracing an ethic of accumulation for higher consumption (Dubois 2012, 113). Of course, it was all-too-convenient for landowners to insist that the responsible exercise of freedom required that the freed people continue to generate immense profits for them. Yet in their argument they did not appeal to their own naked self-interest. Rather, they attempted to make a moral argument.

To most whites, the resistance of freed blacks to wage labor, their reduction in work-hours, and the priority they gave to subsistence farming over production for the market amounted to a rejection of civilization and a reversion to barbarism. Their conduct was taken as proof that blacks were innately lazy, lacking in ambition and a work ethic, and impervious to market-based rational incentives. Abolitionists had argued that the plantation system would survive emancipation. They appealed to Adam Smith, who argued that free labor was more productive than slave labor because people who were rewarded with the fruits of their labor, and thus could improve their prospects through their own efforts, would work harder than slaves (Smith 1904 [1776], I.viii.41–4). (They overlooked his argument that self-employed workers were more productive than workers who had to share the fruits of their labor with employers: Smith 1904 [1776], I.viii.48.) Most whites saw the freed people's conduct as vindicating the traditional view that the lower orders would work only if they were coerced by force or necessity, and as supporting racial theories of black inferiority (Holt 2000; Foner 2007).

Whites would have done well to ask those they judged so harshly for their response. Pompée Valentin, baron de Vastey, secretary to King Henri Christophe of Haiti, complained: "How can they be competent to judge of our differences, if they hear only the clamor and declarations of one party, without the reply and just complaints of the others?" (1818, 8). Defending the economic system chosen by the Haitians themselves, Vastey observed that the slaves had been malnourished and emaciated due to the plantation system's exclusive focus on cash crops. Since winning independence the Haitian people had diversified their agriculture, successfully introduced food crops such as corn, barley, oats, and potatoes, and expanded cultivation of bananas. This enabled the Haitian people to feed themselves. The new agricultural system was "fitted to our wants and worthy of a free people" (Vastey 1818, 53–4).

In psychology, the 'fundamental attribution error' refers to the tendency of people to explain others' behavior in terms of innate dispositions instead of circumstantial factors (Fiske 1998, 370). This pervasive cognitive bias is even more notable when the observer is responsible for arranging the circumstances that lead to the behavior in question. Whites' tendency to ignore their own causal role in structuring freed people's incentives, and to attribute their choices instead to dishonorable innate characteristics, was particularly glaring in the agricultural case. Although they claimed that wage labor offered serious prospects for blacks to improve their economic standing through hard work, they went to great lengths to minimize the wages they had to offer field-hands. In Jamaica they kept wages down by importing indentured servants from India. They preached to black men about the pride they should take in earning enough to support their wives and daughters at home, even while forcing black women to work in the fields (Holt 2000, 55–8).

Finally, and perhaps most astonishingly, in accusing freed people of laziness and barbarism, whites condemned in blacks what they held out as a moral ideal for themselves. The point is not simply that no free white would accept the terms of labor that planters offered the freed people. To be one's own boss, to stake out a homestead and make it one's own by farming it, was the essence of the free-labor ideal that lay at the core of the ideology of the antebellum Republican Party in the United States. Wage labor was merely a stepping stone to independent proprietorship; failure to take that step indicated a "dependent nature", in Lincoln's view (Foner 1995, Kindle loc. 325).

Whites' test of emancipation was thus marred by profound cognitive bias and contradiction. The problem is more profound than the fundamental attribution error. It goes to the core of moral epistemology, of the source of our awareness of moral requirements. Children develop their notions of goodness from their own experiences in pursuing and attaining what they like. They could never acquire any notion of moral right or duty from such experience. Such ideas invariably come from outside the self, from recognizing the authoritative demands of others. The child wants to pick the pretty flowers because they seem good to him, and stops only when his parent tells him they are someone else's property and it is wrong to steal (Dewey and Tufts 1981, 215). The experience of being held to account by another with the authority to do so is indispensable for learning the difference between what is good from one's own perspective and what is morally right.

Suppose a person not only has no one holding him to account, but has the power to enforce the demands he makes on others. Such an experience of unaccountable power would produce profound moral confusion. It would be difficult for such a person to distinguish what is good from his perspective from what is morally right. His power to make others obey his will would make him think that whatever he thought good was obligatory to others (Dewey and Tufts 1981, 226). He would define their moral virtue to consist in their disposition to willingly serve his interests.

Suppose he lost the power to impose his will on others, but the others failed to acquire enough power to hold him accountable for the way he treats them. They would refuse to become mere instruments of his will. Yet they would be unable to get him to acknowledge their right to refuse. He would then judge their resistance to his designs as a moral failure on their part, a demonstration that they were unfit for freedom.

The same psychological mechanism that produces moral confusion in powerful individuals produces it in powerful groups. Since the advent of racialized slavery, whites have defined the virtue of blacks in terms of their service to white interests. Their narcissistic definition has persisted in post-emancipation societies. The stereotype of blacks as lazy is rooted in the demand of whites that blacks work for them, for lower pay and under worse conditions than any white person will accept, and the rejection by blacks of that demand without their having the power to make whites recognize the double standard behind it.

The pragmatist idea that the case for emancipation had to be tested by its actual consequences, and not only by our reactions to it in a thought experiment, was broadly accepted in nineteenth-century Euro-American moral discourse. This idea was correct, but the test was biased. Had the same test of self-directed labor been applied to freed people as to whites, the demonstration that wherever their efforts were not blocked by whites freed people managed to lift themselves out of slavery to independent self-sufficient farming would have counted, as Vastey argued, as a resounding success for emancipation.

5. History, Memory, and Moral Epistemology

Social groups draw moral lessons from their histories. They are right to see that their experiences in trying to live in accordance with certain moral principles provide critical evidence for or against those principles. However, social groups do not always draw the right lessons from their histories. Such failures may be due to morally biased or contradictory tests of success. They may also be due to biased memory and inadequate causal analysis. Since Haiti gained independence, the United States and Europe have repeatedly pointed to Haiti's failure to secure political stability as evidence of blacks' inability to govern themselves. They rarely acknowledged their own roles in fostering political and economic crises in Haiti, through such means as extorted reparations payments, gunboat diplomacy, Western-backed military coups, credit crises engineered by monopolistic banks under French or US control, and military occupation (Dubois 2012). As Vastey complained, "our faults have given strength to the unfavorable disposition of our enemies, and hardened them in their odious prejudices. They are unwilling to ascertain the source of these faults, of which they are the first cause" (Vastey 1818, 37). If we correct such errors, history can point the way beyond first-order assessment of particular moral principles to a more general account of how to improve our moral principles. It can thereby contribute to a naturalized moral epistemology. Such a moral epistemology necessarily focuses on the social organization of moral inquiry, because moral awareness arises from outside the agent, from the claims of others. Moral norms regulate social relationships and facilitate or constrain the possibilities for progress in moral knowledge.

On a naturalized, pragmatist view of moral inquiry, we do not already possess an independent standard of moral rightness against which we can measure the moral success or failure of any particular society's norms. Nor do we model moral inquiry as best undertaken through thought experiments that can be carried out by an isolated individual, or by a demographically narrow sector of society, discussing matters around a seminar table (or in a legislative assembly or executive committee)—particularly not if that sector enjoys relative power and privilege over those affected by or subject to the moral norms under discussion.

Although some things can be learned by these kinds of reflection, we must also be mindful of the profound biases that tend to corrupt the moral reflections of the relatively powerful, when they engage in unaccountable moral inquiry that is, implicitly or explicitly, authoritarian in its social organization.

I have sketched an alternative naturalized approach to moral inquiry in this chapter. On this view, social groups learn to improve their moral norms through historical processes of contention over them. "Contention" encompasses a broad range of activities that may change over time (Tilly 1993). While it includes "pure" moral argumentation, it also includes a variety of other ways of making interpersonal claims, including petitions, hearings, testimonials, election campaigns, voting, bargaining, litigation, demonstrations, strikes, disobedience, and rebellion. We can model the epistemic value of different modes of contention in terms of their potential for inducing error-correction, counteracting bias, clearing up confusion, taking up morally relevant information, making people receptive to admitting mistakes, drawing logical conclusions, and other epistemic improvements. While we do not already have on hand a final standard of moral rightness, we may have a fairly good idea of characteristic sources of moral error, ignorance, bias, and blindness, drawn from social and cognitive psychology and from historical investigation. In different social contexts, different modes of contention may be required to overcome these sources of bias, to open people's minds to morally relevant considerations, and their conduct to moral accountability.

A major source of bias is unaccountable power over others. It is extraordinarily difficult for social groups that exercise unaccountable power over other groups to distinguish what they want subordinate groups to do for them from what those groups are obligated to do. It is extraordinarily difficult for dominant groups to recognize when they are behaving unjustly toward subordinate groups. Power makes people morally blind. It stunts their moral imaginations and corrupts their moral reasoning, tripping them up in contradictions and sophistries. Successful contention by subordinate groups, in historic moments where they are in a position to make themselves heard and hold dominant groups accountable, sometimes breaks through the vanity, smugness, and narcissism of the powerful, as in 1794, when the National Convention abolished slavery throughout the French empire.

At such moments, we see glimpses of a democratic as opposed to an authoritarian organization of moral inquiry. "Democratic" in this context does not mean that the right is determined by majority voting. It means that all sides to a moral dispute—those making claims, those to whom the claims are addressed, those otherwise affected by claim-making—manage to participate on terms of equality in contention over the principles governing their claims, and do so in ways the others cannot ignore or dismiss but must address in their own terms. Moral knowledge comes from outside, not inside the self. It requires openness to the claims and perspectives of others.

Democratic inquiry does not solve all problems in moral epistemology. Nor is contention the only path to moral insight. Sometimes the powerful can be stirred into recognition of the full humanity of subordinates through intimate association on terms of equality. This is a common pattern among feminist men. Condorcet found his intellectual soulmate in Sophie de Grouchy, John Stuart Mill in Harriet Taylor, William Thompson in Anna Doyle Wheeler. More generally, friendly or cooperative association across identity-group boundaries is key to prejudice reduction (Allport 1954), which checks a major source of moral error. In the absence of intimacy on terms of equality, people can be stirred to sympathetic moral recognition of others through autobiography, journalism, fiction, drama, painting, and other arts. Here too, the key to moral insight is receptiveness to others in their full humanity.

Every story we tell about how groups' moral convictions have changed implies a background moral epistemology. Time and again in the history of moral progress, the oppressed have taught moral lessons to the powerful. Time and again, the historical memories of dominant groups erase those events and replace them with an imagined rational reconstruction of the acquisition of moral insight through the self-sufficient reasoning of the dominant. What countries took the lead in insisting on the legal application of the 1948 Universal Declaration of Human Rights to all human beings? The United Kingdom and France, perhaps, with their long human-rights traditions? No. Both countries argued vigorously for a "colonial clause" that would exclude colonial subjects from claiming the rights that the UDHR said all humans were entitled to simply because they are human. The United States? No. The United States joined England and France on the colonial clause, in return for their support of a "federal state clause", which would exempt the member states of any federal government from being subject to the law. The United States wanted to assure its southern states that ratification of the UDHR and its associated legally enforceable covenants would leave their systems of white supremacy intact. The countries that took the lead in insisting that the UDHR was really universal were former colonies of Europe and the United States, notably including India, the Philippines, and Panama (Roberts, 2014). We forget such histories at our moral peril, for progress in moral inquiry requires the practice of epistemic justice by and for all.

5

Group Emotion and Group Understanding

Michael S. Brady

Introduction

It is a commonplace that our emotions can lead us astray, in action and in belief: Jane's fury at being overlooked for promotion might cause her to punch her boss; Joe's fear of the dark could convince him that there are monsters under the bed. But if individual emotional experiences have a bad reputation, group or collective emotion can often seem even worse. Partly this is due to the fact that group emotion can generate greater disvalue or evil than that typically caused by individual emotion—as illustrated by the Salem witch trials, stock-market runs, or football hooliganism. Partly this is because the group nature of the emotion generates attitudes and behaviours that are 'out of character', in the sense that they are states and actions that the individual wouldn't have and wouldn't perform without the influence of the group—as was the case with many individuals caught up in the public outpouring of grief when Princess Diana died.

It would, therefore, be a mistake to deny that sometimes individual and group emotions merit criticism along both epistemic and practical dimensions. Nevertheless, focusing on the negative outcomes of individual and group emotion should not blind us to the positive value that individual and group emotion can clearly have. It is obvious that individual emotions can have epistemic value in so far as they are apt or appropriate: think of the resentment a university lecturer might feel if the senior management award themselves a 10 per cent pay-rise while academic staff are forced to take a pay-cut. It is equally obvious that group emotion can have epistemic value in virtue of being apt or fitting too: think of the public anger when some MPs were filing bogus expenses claims, or the public pride and joy experienced in the United Kingdom during the 2012 Olympics.

In this chapter I want to make the case for the epistemic importance of group emotion along other dimensions, which are rather less obvious and which have been little discussed. In particular, I want to explain how group emotion can help to bring about the highest epistemic group good, namely *group understanding*. Moreover, I will argue that this group good would be difficult to achieve, in very many cases, in the absence of group emotion. Even if group emotion sometimes—indeed often—leads us astray, we would be worse off, from the standpoint of achieving the highest epistemic good, without it. The structure of the chapter is as follows. In section 1, I present a schematic account of the elements of individual and group emotion. In section 2, I explain how individual emotion plays an important role in enabling individuals to understand their evaluative situation. And in section 3, I argue that this provides a model of how group emotion can promote group understanding in an analogous way.

1. Individual and Group Emotion

What is an emotion? There is—and this is hardly a surprise—a lack of consensus on the necessary and sufficient conditions for something to be an emotion. But there is widespread agreement that paradigm cases of emotion involve a number of elements; rival accounts of emotion diverge, typically, on the issue of which of these elements is most important, or has explanatory priority. An example will help to illustrate this idea. Suppose that I've been interviewed for my dream job, and feel very happy about my performance. I form an expectation that I will be offered the job, given how well suited I am to the position, and how well I think that I performed at interview. The chair of the hiring committee now telephones to tell me that the job has been offered to someone else. Upon hearing this, I feel an immense and crushing disappointment. My emotional reaction has, it seems, certain 'parts' or 'elements'. These are: (i) a perceptual experience, in this case an auditory one, as of the chair telling me that I didn't get the job; (ii) a belief, following quickly and automatically from the auditory experience, that I didn't get the job; (iii) an appraisal or evaluation that this is a bad thing to happen, made worse because I had high hopes and expected to be offered the position; (iv) facial and bodily changes: my shoulders slump, my stomach lurches, I frown, I am on the verge of tears; (v) feelings, in this case an experience of negative affect or valence, which is (perhaps in part, perhaps in whole) an experience of the body and facial changes described above; (vi) motivational or action tendencies, such as to throw the phone across the room, to scream and shout, to head to the pub to drown my sorrows; (vii) cognitive changes: I focus on what I might have done wrong, I imagine telling my family that I didn't get the job, I think about the embarrassment of facing colleagues on Monday morning, and so on. A further element (viii), though not one that is best understood as part of the emotional experience, is widely agreed to be necessary for emotional experience; this is some underlying care and concern, in virtue of which the emotional reaction or response makes sense. In this instance, the additional element is a strong desire on my part to get the job, or some other care or concern that getting the job would ultimately satisfy.

Although almost everyone holds that these elements are involved in paradigmatic cases of emotional experience, there is, as noted, significant disagreement over which are essential, over the temporal and causal relations between the elements, and over relations of normative and explanatory priority between them. It is no part of my remit here to address, let alone settle, such disagreement. But I do want to note the obvious point that there must be some relations between these elements if we are entitled to regard them as parts of the *same* emotional experience. So at the very least, the auditory perception is a causal precursor to the belief that they were offering the job to someone else, which is a precursor to an evaluation of this as a bad thing. Perhaps the bodily, affective, motivational, and cognitive changes occur after we have evaluated the situation as bad, or perhaps some of them are prior to or simultaneous with this; but these too will be related to each other on whatever story we tell. For instance, we might claim that our attention is drawn to the event, or its badness, by the affective phenomenology of the experience; or we might claim that this phenomenology just is an experience of the behavioural imperative that is generated by an appraisal of the event as bad; and so forth. Whatever the correct account turns out to be, we can make this minimal claim: the elements of emotional experience enjoy causal and normative relations with each other, such that it makes sense to regard them as part of the same experience or the same mental state.

What, then, are group emotions? We might try to answer this question by providing accounts of group counterparts of the elements found in paradigmatic cases of individual emotion—that is, perception, belief, evaluation, facial and bodily changes, feelings, action-tendencies, and underlying concerns—and then saying something about how these group states are linked to form group emotion. But this strategy faces significant difficulties. Firstly, we might wonder whether there can be *genuinely* group mental states, or at best only an aggregate of individual mental states. Is there such a thing as genuinely group belief, for instance, or only an aggregate of individual beliefs? The task of answering this question is made more difficult when we recognize that there are significant differences between the kinds of things we identify as groups or collectives: a family, co-workers, followers of a religion, a scientific research team, an institutional committee, a political party, a book club, a nation, a sporting crowd, hobbyists, lovers, an online gaming community, the rebel alliance. Isn't it possible, indeed likely, that the group beliefs of lovers will be very different, in metaphysical kind as well as propositional content, from the group beliefs of a political party, an online community, or a team of scientists?

Even if there are genuine group beliefs, and even if we apply this term to the same kind of state across different groups or collectives, we can nevertheless doubt that *some* elements of individual emotional experience have genuine group or collective counterparts: there are no such things as group bodily and facial changes, for instance; 'the body politic' and 'the face of the company' are clearly metaphorical uses. And even if we can make the case for genuine group beliefs and intentions, it will presumably be harder to make the case for genuine group feelings and memories. As a result, the prospects for this kind of answer to our question about the nature of group emotion seem dim.

I propose that we adopt another strategy, which is to employ a model of group emotion that emphasizes the links or connections between individual emotions, and remains (relatively) silent about the nature of group counterparts of belief, appraisal, bodily changes, action tendencies, and other elements of emotions as experienced by a single subject. In doing so, I am not just tailoring my account of group emotion so that it fits the positive case I want to make. For the account of group emotion I will employ is one that appears best fitted to illustrate the *negative* epistemological effects that group emotion can have. As a result, I want to make a case for the epistemic credentials of group emotion as it might be understood by those who are sceptical as to its epistemic worth.

The picture of group emotion I will work with starts from something that ought to be acceptable to all, namely the view that group emotion involves or is partly constituted by individual emotions.[1] To illustrate, consider the student protests in London in 2011 over the Government's proposal to increase tuition fees. This is, plausibly, an instance of group anger. And at the very least, this group anger involves, or is partly constituted by, the individual anger that each student feels towards the decision about fee increases. Now the individual students in the group would each, typically, believe that fees are going to increase, appraise this as a bad thing, be motivated to respond in an appropriate way,

---

[1] In thinking about group or collective emotion I have benefitted greatly from reading the following works: Gilbert (2001); Schmid (2009); Salmela (2012). The thought that group emotion requires 'synchronization' between individual emotions, and that part of this process involves a desire for 'affective conformity', are due to Salmela and Schmid, respectively.

experience bodily and facial changes as a feeling of anger, pay attention to the Government's proposal, and be concerned for their future studies. Of course, this is rather simplified, and we can be sure that there are significant differences between the students' experience of anger on a number of dimensions. Nevertheless, we might think that it makes sense to talk about the group anger over the fee increases only in so far as we have (enough) individuals who are undergoing individual emotional experiences of roughly this type.

However, a similarity or commonality in individual emotions to some event does not make this a group emotion. So the mere fact that many individual students feel anger towards the Government's decision does not suffice for there to be a group emotion. At the very least, individuals have to be *aware* that others are feeling as they are feeling, in order for there to be the possibility of group emotion. But beyond this, two further connections are important and prominent—especially in the kinds of group emotion often criticized for leading people astray. One is involved in the generation of new cases of individual emotion via 'emotional contagion', which is 'the tendency to automatically mimic and synchronize expressions, vocalizations, postures, and movements with those of another person and, consequently, to converge emotionally'.[2] Thus I might come to grieve at the death of Princess Diana as a result of becoming aware that others are grieving; their grief 'transfers' itself to me, in an automatic and non-conscious way. As a result of contagion, a certain emotion is conveyed to and spreads throughout some group; *we* come to grieve as a result of this process of mimicry and synchronization.

A second kind of connection, which is plausibly involved in cases where individuals who are already experiencing some emotion encounter others who are feeling the same way, is a form of acceptance and endorsement of the emotions of others. Thus *we* might come to be angry at the Principal's pay rise when we become mutually aware of the individual anger of others directed towards this event, and mutually accept and endorse the fact that others feel as we do. This need not involve anything like explicit endorsement or acceptance: we can welcome the fact that others are feeling in the same way without explicitly acknowledging that this is what we are doing, to ourselves or to others. Endorsement doesn't require reflection and deliberation resulting in anything like a *decision* to endorse how others feel. Instead, it is plausible to assume that the synchronization of individual emotions occurs as a result of a desire for what Hans Bernard Schmid terms 'affective conformity': the thought is 'that people *enjoy* being in the same affective state as those around them, independently of the

---

[2] Hatfield, Cacioppo, and Rapson (1994), 5.

mode of the feeling at stake. That's why sharing increases the joy, but diminishes the pain.'[3] In this way we welcome the fact that others are feeling as we do and identify our feeling with theirs, and this process of acknowledgement results in the convergence of our emotions with the emotions of those around us, such that it seems appropriate to refer to the resulting state as one of group or collective emotion.

This process of mutual awareness, emotional contagion, and affective conformity might also plausibly be viewed as resulting in group counterparts of the elements of individual emotional experience. Thus, the process of mutual awareness will result from (or indeed consist in) the individual attention of each person being focused both on some emotional object or event, and also focused on the fact that others are attending to this object or event. In this way we might talk about *our* attention being focused on the object or event in question: when we become mutually aware that we are all individually sad *about Princess Diana's death*, then we might say that Princess Diana's death becomes the object of group attention. By the same token, mutual awareness that others are undergoing a similar emotional experience involves awareness that others make similar appraisals of, have similar feelings about, and are similarly inclined to act with respect to the relevant object or event. In so far as we accept and endorse these similarities, we might talk about *our* appraisal of, feelings about, and behavioural tendencies towards such an object or event. So the process by which individual emotions are synchronized and converge can be regarded as a process in which the group comes to attend, appraise, feel, and be inclined towards action.

This is, of course, merely a sketch or outline of one model of group emotion. But it does seem to capture a good deal of what people mean when they refer to group emotion. Moreover, by emphasizing the processes of emotional contagion and affective conformity, the model seems to include the kinds of non-rational and non-conscious processes that many will regard as having deleterious effects on our epistemic lives. For we ought not, other things being equal, to have emotional states because other people are in those states and because we want there to be conformity between ourselves and others. Emotions, after all, are representational states involving appraisals of our environment; and representational states ought to be responsive to evidence rather than desire. The model should therefore be acceptable to those who are sceptical as to the practical and epistemic benefits of group emotion. As a result, I will assume that something like this model of group emotion is correct, and proceed to argue that group emotion, understood in this way, is extremely important, if not essential, for the production of the highest group epistemic good. In the next section I'll begin this argument by

---

[3] Schmid (2009), 66.

showing how individual emotion is essential for providing us with individual understanding, and, in section 3, I will make an analogous case for group emotion and group understanding.

2. Individual Emotion and Understanding

I have argued elsewhere that individual emotions can have significant epistemic value under the right conditions; and one important way in which they do so is in the promotion of our *understanding* of our world and of ourselves.[4] Central to this argument is the fact that emotion and attention are very closely linked. One aspect of this relationship is that emotions can make things *salient* for us, or can alert us to potentially important objects and events. For example, my anxiety over my new white carpet makes salient the fact that you're drinking red wine; my fear when camping in the woods draws my attention to all of the sounds outwith the tent; my delight at being reunited with my beloved makes salient all of the wonderful things about her; and so on for many other cases of emotion.

However, emotions such as fear and joy do not just automatically and reflexively direct and focus attention; one of the other things that emotions tend to do is to *capture* and *consume* attention. To say that attention is captured and consumed by emotional objects and events is to say that such objects and events hold sway over us, often making it difficult for us to disengage our attention and shift focus elsewhere. Think, for instance, about what it is like when one is awoken, in the dead of night, by a noise outside of the tent and experience fear. In normal circumstances one's fear is *not* over very quickly; rather, one remains in a fearful state as one listens attentively for further noises, tries to think of possible non-threatening explanations, rehearses strategies for dealing with the potential danger, considers possible escape routes, and so on. Similar points can be made about jealousy, anger, resentment, sadness, shame, guilt, love, and many other emotions.

Now the reflexive and automatic focusing of attention in emotional experience can enable us to quickly and efficiently notice things that are important for us to notice, and so can have epistemic value along this dimension. Might attentional consumption have a similarly valuable role to play in our emotional lives? I think it does. For one of the important things that attentional persistence can do is to

---

[4] In Brady (2013). The caveat about conditions is of course important; clearly I don't want to argue that emotions always make us better off from the practical or epistemic standpoint. But I do want to say that without emotions it would be difficult for us to attain the most valuable epistemic good of understanding, and that it is to this extent that emotions have significant epistemic value. The reasons for thinking emotions important in this way will be explored in what follows.

enhance our representation of potentially significant objects and events, by enabling us to discover reasons which bear on the accuracy of our initial emotional appraisals.[5] In other words, the persistence of attention can facilitate, by motivating the search for and discovery of reasons, a judgement as to whether emotional appearance in this instance matches evaluative reality.[6] If this is correct, then emotions involve two important links with attention: they can direct and focus attention, thereby alerting us to the presence of potentially important or significant objects in our environment; and they can capture and consume attention, thereby enabling us to determine whether things are as they emotionally appear.

Note that there is a need for reassessment, and for a number of reasons. One is that 'fast and frugal' emotional responses can be relatively indiscriminate. In other words, the appraisals involved in many emotional responses are 'quick and dirty': very rapid but relatively coarse responses to emotional objects and events. This is why people are reflexively, automatically, and indiscriminately afraid of things like crawling insects, loud noises, looming objects, and so on, only some of which will actually be dangerous.[7] So discrimination is the price that has to be paid for ensuring speed of response; it is, nevertheless, a price that is worth paying in certain circumstances, given that it is better to have an emotional system that responds very quickly to all such things than it is to have a more discriminating evaluative system that responds more slowly.[8]

Another reason is that, as Justin D'Arms and Daniel Jacobsen have pointed out, 'emotions involve powerful motivational tendencies, so regulating them is an indirect way of regulating behaviour'.[9] As D'Arms puts things elsewhere:

states such as anger, envy, and shame, for instance, involve motivational tendencies: toward retaliation, competition, or concealment, respectively. Because it matters very much to each of us how we act, there's reason to think about what to be angry, envious or

---

[5] See de Sousa (1988), 196. I claim that this is one of the things that attentional persistence can do, because I think that it can have other roles or functions as well, in particular a monitoring function of keeping the object in view so as to promote the correct behavioural response.

[6] See Lee Anna Clark and David Watson, who write that, 'triggered by environmental events, emotions act as salient internal stimuli that alert the organism to *the need for further information gathering* and action': Clark and Watson (1994), 131 (my italics.) On my view, this need is best served through the consumption of attention in emotional experience.

[7] See e.g. Barrett (2005).

[8] As Phoebe Ellsworth writes, 'one of the central functions of emotion is to motivate the organism to respond quickly and effectively to environmental threats as they arise. Generally the costs of failing to respond soon enough are far greater than the costs of responding when it is not really necessary... It is far safer for an organism to be calibrated to feel emotion when it is not warranted— to have a hypersensitive system—than it is to have a system that postpones the initiation of emotional processes until there is no question that they are justified.' Ellsworth (1994), 194.

[9] In D'Arms and Jacobsen (2006), 99–126.

ashamed of. These facts generate an important role for intrapersonal criticism and reflection that an agent can undertake concerning the appropriateness of these irruptive, motivating emotions.[10]

In so far as emotions facilitate such reflection on the appraisals that partly constitute the emotional responses themselves, through the consumption of attention on to the objects of those responses, then emotions have value in enabling us to achieve a more discriminating response, and in allowing us to control and regulate our emotional take on the world.

There is, moreover, considerable evidence that emotions do indeed promote such reflection and reappraisal, and that part of their epistemic value precisely consists in the fact that reflection and reappraisal will often not occur—or will be more costly and less effective—in the absence of emotion. This is evidence that *reason alone* is ill-suited to the important task of facilitating more accurate emotional appraisals. Phenomenological support for the idea that emotions facilitate reappraisal, through effects on attention, is common: we often *feel the need* to discover reasons and evidence—when awoken by a strange noise, we seek out evidence to confirm (or hopefully disconfirm) our initial assessment that we are in danger. When jealous we feel a motivation to seek confirming (or hopefully disconfirming) evidence of infidelity. So a need to discover reasons is often felt as emotion persists. Moreover, it also seems true that when we are no longer emotional we usually *lack* the motivation to check or assess the accuracy of our initial emotional appraisals. If I no longer feel afraid, then it is unlikely that I'll bother myself much with seeking evidence as to whether or not I'm in danger. If I no longer feel jealous, it is doubtful whether my attention will remain fixed on the prospect of infidelity, or that I will expend effort in an attempt to determine whether or not my partner really is unfaithful.

The phenomenological evidence fits in nicely with views in psychology which suggest that appraisal and reappraisal is an ongoing process in emotional experience. Thus, Klaus Scherer has argued that 'emotion *decouples* stimulus and response', allowing a 'latency period between stimulus evaluation and reaction'.[11] On his view, 'the first major function [of the latency period] is the ongoing analysis of the stimulus event, which allows the organism to arrive at a more detailed or more realistic conclusion and may lead to a re-evaluation and consequently a revision of the original appraisal'.[12] In the same vein, Richard Lazarus writes:

[f]or people to react with an emotion, the relevance of what is happening to their well-being must be sensed, as well as whether this has negative or positive implications. We do

---

[10] D'Arms (2005), 8.　　[11] Scherer (1994), 128.　　[12] Scherer (1994), 129.

not stop with a hasty and incomplete cognitive evaluation—this constitutes an incompleted task, which the person or animal is compelled to pursue further—until what is happening can be understood in a way that is relevant to efforts at coping. Although the initial appraisal may be hasty and limited, if the opportunity for further investigation of what is happening presents itself, it would be a strange creature that let things drop before a full functional understanding has been achieved.[13]

If Lazarus is right, the compulsion or motivation to investigate further and to attain a more accurate appraisal is the norm: it is not just that we sometimes feel the need to investigate when we have the opportunity, but rather that feeling the need to investigate when we have the chance is what normal humans do.[14]

The idea that emotion plays the role of facilitating reflection and reappraisal through the capture of attention, and that reason alone is not up to this task, also finds support in the writings of Thomas Reid, who is one of the few philosophers to be concerned with the connection between emotion and attention. Reid claims that '[i]t requires a strong degree of curiosity, or some more important passion, to give us that interest in an object which is necessary to our giving attention to it. And, without attention, we can form no true and stable judgement of any object.'[15] And: '[a]ttention may be given to any object, either of sense or of intellect, in order to form a distinct notion of it, or to discover its nature, its attributes, or its relations and so great is the effect of attention, that, without it, it is impossible to acquire or retain a distinct notion of any object of thought.'[16] So for Reid, emotion (or 'passion') is necessary for us to pay attention to some object or event, and paying attention is necessary for us to form an accurate ('a true and stable') judgement about that object or event. Reid would therefore be sympathetic to the idea that emotional control of attention facilitates a better grasp of our evaluative situation, by making us aware of the reasons that have a bearing in these circumstances. Now although Reid's claims about necessity are too strong, since we can intentionally fix our attention onto some object in the absence of emotion, he is nevertheless surely correct to stress the importance of emotion in the direction and control of attention, and the importance of attention to an accurate evaluation of our circumstances. Although we *can* intentionally fix and direct our attention onto some object or event, this is usually very costly in terms of mental resources, in which case there is a significant advantage in having a system which keeps our attention fixed with little in the way of conscious effort on our part. If considerations of mental economy speak in favour of the automatic

---

[13] Lazarus (1994), 215.

[14] There is, moreover, considerable neurophysiological evidence for the view that attentional persistence promotes enhanced representation of emotional stimuli. See e.g. LeDoux (1996).

[15] Reid (1969), 184–5.   [16] Reid (1969), 76–7.

and reflexive direction and focus of attention in emotional experience, then similar considerations speak to the emotional consumption of attention. So even if the emotional consumption of attention is not strictly necessary for us to get an accurate picture of our evaluative situation, the emotional governance of attention for this end is extremely valuable.

How, then, is all of this related to the epistemic goal of understanding? The simple answer is that the search for and discovery of reasons that bear on the accuracy of our (initial) emotional response is a search for features that constitute reasons as to why some object or event has an evaluative property like 'dangerous' or 'shameful'. But then the reassessment of our initial emotional appraisal will involve trying to *understand* one's evaluative situation: for an attempt to discover why an object is dangerous or why it is harmless just is an attempt to achieve an evaluative understanding of the relevant object or event. Understanding, after all, involves a grasp or awareness of the connections or links between various items. It involves seeing how things fit together, how features are related, how facts support and explain other facts. Discovery of the danger-making features of some object or event is a discovery of how these features support and explain another feature, namely dangerousness; we thus make sense of the dangerousness of the object or event when we come to grasp the reasons why it is dangerous, rather than simply coming to grasp *that* the object is dangerous. As a result, the search for and discovery of reasons that emotional experience facilitates is a search for an accurate understanding of value; and when this search is successful, and we grasp why the object is dangerous (if it is), we do not just attain a more accurate emotional judgement about our situation—although this is certainly of epistemic importance and value. We also attain an understanding of our evaluative environment. Indeed, given the importance of emotion in motivating the search for reasons, we can doubt that the epistemic goal of understanding would be (easily) achievable in the absence of emotion. If this is correct, then emotions are of considerable importance in enabling us to achieve one of the highest epistemic goods.

In the following section I'll show how this account can be extended, and will argue that group emotion has significant epistemic value in facilitating group understanding of important objects and events.

3. Group Emotion and Group Understanding

I want to argue that group emotion can parallel individual emotion in having epistemic value along two dimensions. First, group emotion can draw attention to some important or significant event; second, it can motivate or facilitate group understanding of that event. In particular, I will focus on one important way in which group emotions generated by emotional contagion and affective conformity can effectively bring about these epistemic goods, by drawing the attention of those in power to the importance of some event, and by motivating *them* to arrive at and make available to the group an understanding of that event. The thought that group emotion can have epistemic value in this way can be illustrated if we focus on the institution of the *public inquiry*.

The idea that there is a connection between group emotion and public inquiry is apparent from even a brief survey of newspapers and websites. For instance, in June 2013 the *Financial Times* reported that 'Dublin aims to set up a public inquiry into the banking crash in the autumn in response to deep public anger caused by the broadcast of taped phone calls suggesting Anglo Irish Bank deliberately misled the previous government into supporting the lender during the financial crisis'.[17] This is a case where group emotion *generates* a public inquiry, as a result of the recognition of such anger by a governing body, and a desire of that body to address this anger through setting up an inquiry. There are very many cases which display the same structure: in 2011, public outrage at phone-hacking by employees of News International was a major factor in motivating the Levenson Inquiry; anger at the spiralling costs of the Scottish Parliament Building motivated the Fraser Inquiry in 2004. Indeed, the UK Government's own list of public inquiries inclines one to believe that group or public emotion (typically anger, in its various forms) was a motivating factor in generating most if not all of the recent cases where an inquiry has been set up.[18]

Why is there such a close connection between group or public emotion and the setting up of public inquiries?[19] A central reason is that public inquiries are held with respect to events about which there is group or public concern.[20] It is plausible to maintain, in light of this, that group emotion can play the epistemic

---

[17] *Financial Times* (2013).

[18] See e.g. the list at <http://www.nationalarchives.gov.uk/webarchive/public-inquiries-inquests.htm>.

[19] I talk of group *or* public anger here merely to reflect that sometimes the emotion will be that of a particularly well-defined group of citizens, and sometimes the emotion will be shared by a much larger group so that it is perhaps appropriate to talk about emotion that is felt by the public at large. The latter case is still a form of group emotion, if there is common public knowledge that others share the emotion. I'd like to thank a reviewer for this volume for pushing me to be clearer on this point.

[20] At least, this is the case according to the UK Government's Enquiries Act, 2005, §1: 'A Minister may cause an inquiry to be held under this Act in relation to a case where it appears to him that—(a) particular events have caused, or are capable of causing, public concern, or (b) there is public concern that particular events may have occurred.' <http://www.legislation.gov.uk/ukpga/2005/12/section/1>.

role of alerting or drawing the attention of *the governing body* to the fact that some event *is* of considerable public concern. In other words, group emotions can let the governing body know that something is of concern or importance to them, and hence is a proper subject or target for a public inquiry. Just as individual emotion can focus an individual's attention on to some important object or event, so too can group emotion draw a governing body's attention to some event that is of significance to the group. Moreover, and to mirror arguments made earlier, in the absence of group emotion to draw attention to the event, it is highly unlikely that the governing body will recognize it as an event that is of concern to the group, and hence as one that might merit a public inquiry. For if the public do not react emotionally to the event, this is good evidence that the event fails to impinge upon something that matters to them, and hence good evidence that it is not an event of public concern. Group emotion might therefore be regarded as very important to, if not essential to, the generation of public inquiries set up to address and respond to public concerns.

What, then, of the link between public inquiries that are generated by group emotion, and group understanding? Here the connection is straightforward: for one of the central aims of public inquiries is, precisely, to arrive at and to make available to the some public group an understanding of the event that is of concern to that group.[21] As a result, group emotion can promote group understanding by generating inquires that aim at and, when successful, achieve an understanding of some event, where this understanding is then conveyed to the group via a public report.

This is supported both by a general account of the role of public inquiries, and by particular examples of such. Lord Laming, who chaired the Victoria Climbié Inquiry in 2001, proposed the following account of the general aims of public inquiries. On his view, inquiries:

provide an assurance that the facts surrounding an alleged failure will be subjected to objective scrutiny. *They are expected to reach judgements on why terrible events happened.* They often make recommendations on how such events might be prevented in future. They may give relief to some and allow the expression of anger and outrage to others. They are often disturbing and painful events. *They should improve our understanding of*

---

[21] Of course, there are other things that both the group emotion and the public inquiry aim at: holding the guilty to account, enacting changes in procedures and laws, reassuring the public, and the like. But none of these would seem to conflict with the goal of understanding, and indeed, it is plausible that successful achievement of these is more likely if understanding has been achieved. So recognition that there are other goals is clearly compatible with my thesis about the epistemic value of collective emotion.

complex issues. At best they change attitudes, policies and practice. That being so they occupy an important place in our society.[22]

I've highlighted two places where the idea that the role of public inquiries is to promote understanding is explicitly stated. It is not implausible to hold that Lord Laming's statement reflects a widely held view about the role and function of public inquiries: of what public inquiries are and what public inquiries do.

Let us turn now to a particular example. In 2013 the Francis Report was published. This was the result of a public inquiry into the failings of the Mid Staffordshire NHS Foundation Trust. The Report points out that, 'between 2005 and 2008 conditions of appalling care were able to flourish in the main hospital serving the people of Stafford [in the UK] and its surrounding area'.[23] Basic care for patients in the hospital was severely lacking, and the mortality rate was exceptionally high; estimates are that between 500 and 1,200 additional people died there in the three-year period in comparison to NHS mortality figures for the rest of the United Kingdom. This generated significant public alarm. Now, as Robert Francis's 'Introduction' to the Report states, these failings were uncovered in the main 'because of the persistent complaints made by a very determined group of patients and those close to them. *This group wanted to know why they and their loved ones had been failed so badly*.'[24] The Inquiry was set up, at least in part, to provide such understanding. This reading is supported elsewhere in the Report. Francis notes that in 2010 the Health Secretary, Andrew Lansley, explaining his decision to hold a public inquiry, told the House of Commons: 'So why another inquiry? We know only too well every harrowing detail of what happened at Mid Staffordshire and the failings of the trust, *but we are still little closer to understanding how that was allowed to happen by the wider system*. The families of those patients who suffered so dreadfully deserve to know, and so too does every NHS patient in this country.'[25] The failings of the Trust were common and public knowledge; so the public inquiry was not set up to tell us that there *were* failings. Instead, given this knowledge, the inquiry aimed at providing

---

[22] Lord Laming (2004) (emphasis mine).   [23] Francis (2003), 13.
[24] Francis (2003), 13 (emphasis mine).
[25] Francis (2003), 15 (emphasis mine). Lansley continues, on p. 16: 'Why did the primary care trust and strategic health authority not see what was happening and intervene earlier? How was the trust able to gain foundation status while clinical standards were so poor? Why did the regulatory bodies not act sooner to investigate a trust whose mortality rates had been significantly higher than the average since 2003 and whose record in dealing with serious complaints was so poor? The public deserve answers.' Here too we see the central idea, namely, that public inquiries aim at answering such questions and providing the public with an understanding of the moral wrong.

the families of patients, and indeed 'every NHS patient in this country', with an understanding of why there were such failings.

If this is correct—and there seems to be a strong correlation between general claims about the aims of public inquiries and the published reports of such inquiries—then we can hold that group emotion has epistemic value in so far as it generates public inquiries that aim at and, when successful, provide group understanding of some important or significant event. It is, moreover, plausible to hold that in many cases inquiries would not be set up and understanding not achieved in the absence of group emotion. Recall, in support, the description in the Francis Report of 'persistent complaints made by a very determined group of patients and those close to them'. It is plausible to suppose that part of what made the patients and those close to them so determined, and part of what motivated them to make persistent complaints, was the persistence of their anger, which kept attention fixed on the terrible things that had happened, which generated the need to find out why these things had happened, and which motivated and coordinated their behaviour so as to alert the governing body to such events. Once again, in the absence of emotion it is difficult to envisage the patients and those close to them being so determined or persistent; that level of focus and determination is difficult to achieve in the absence of emotion. As a result, group emotion would seem very important, if not essential, for the kind of attentional focus and persistence that is needed in order to alert some governing body to an important event, and to keep attention fixed on that event until understanding is achieved and shared.

4. Conclusion

If all of this is correct, then there is reason to be more optimistic about the epistemic value of group emotion than people have tended to be. In particular, we have reason to be more optimistic about group emotion that has been generated by processes—such as emotional contagion and affective conformity—that seem ill-suited to producing states of epistemic worth. For the kinds of group emotion that result from these processes can still play the significant role of alerting governing bodies to the fact that something important is of concern to the public, and motivating the governing bodies to satisfy this concern by promoting an improved public understanding of the events in question.

Of course, none of this supports the idea that group emotion always makes us better off, or never leads us astray. Clearly there can be serious and negative epistemic effects of group emotion. Furthermore, none of this implies that governing bodies set up public inquiries because they have a genuine desire to arrive at an understanding and to convey this to the general public. Perhaps the motives of the governing body are (as is often the case) rather more dubious and self-serving than this. Nevertheless, none of that counts against my general point, which is that there are epistemic goods that we would be very hard pressed to achieve in the absence of group emotion. And perhaps, if we think that the epistemic good of understanding is worth the epistemic dangers and disadvantages that group emotions can bring, we might be inclined to view group emotion in a more favourable light than we have previously.

6

Changing Our Mind

Glen Pettigrove

Introduction

For creatures like us, the journey to knowledge follows an oddly circuitous route. We seldom proceed by steady steps in a single direction. Rather, we follow hunches, make intuitive leaps, wander down blind alleys, backtrack, and suddenly stumble upon a more promising path. So if we are to prove capable of increasing our store of knowledge, one of the things we must learn is to revise our beliefs. The same holds true for the groups of which we are a part. If they are to prove capable of possessing and increasing their knowledge, they too will need to be capable of revising their beliefs. While the nature of collective knowledge has begun to receive more attention in recent years, surprisingly little work has been done on the process of belief revision that makes it possible. And that which has been done has focused on a very particular kind of community and a very particular kind of knowledge, namely the scientific community and scientific knowledge. My aim is to extend the analysis of collective belief revision to a different sort of community and a different sort of knowledge, namely, moral communities and moral knowledge. This chapter will suggest that, whatever one's preferred account of what groups know, we need a richer account of belief revision to support it than has been offered thus far. And it will propose one such alternative.

1. Current Accounts of Collective Knowledge and Collective Belief Revision

Over the past twenty-five years Margaret Gilbert has been developing a distinctive account of collective agency (1989, 1996, 2000, 2013). In the course of so doing she has defended not only claims about collective agents and their actions but also claims about collective beliefs and collective emotions. According to her account of collective belief: 'There is a *collective belief that p* if some persons are jointly committed to believe as a body that *p*' (2000, 39). Persons are jointly committed to a belief if 'each of the parties has expressed his or her personal willingness to be party to it in conditions of common knowledge', that is, in conditions where each of the parties committing to the belief 'has expressed his or her personal willingness to be a party to the joint commitment' (2000, 40). Under such conditions, she argues, it is appropriate to say: 'We believe that *p*.'[1]

Starting from an account like Gilbert's of (a) collective belief that *p*, one might add the familiar conditions of (b) the group's warrant to believe that *p*, and (c) the truth of *p*, in order to generate an account of group knowledge that *p* which parallels accounts of individual knowledge.[2] Of course, one might challenge such an account for reasons that have been well rehearsed in contemporary epistemology. One might worry, for instance, that conditions (a)–(c) fail to distinguish between cases of good epistemic luck of the sort Gettier (1963) highlighted and cases of 'genuine' knowledge. Or one might prefer a virtue-epistemic account of knowledge, according to which: 'Knowledge is a state of cognitive contact with reality arising out of acts of intellectual virtue' (Zagzebski 1996, 270). Or one might challenge the account on the grounds that it fails to be sufficiently attentive to the fact that our ascriptions of knowledge are context sensitive, such that we hold agents to quite different epistemic standards in different conversational contexts (DeRose 2000). More will be said about knowledge and other epistemic states, like understanding, in section 2; however, for the most part the discussion will set these familiar disputes among epistemologists to one side. This is not because they are uninteresting or unimportant. On the contrary, those working on collective epistemology would do well to pay more attention to the implications of the conversational contexts in which we ascribe knowledge to groups.[3]

---

[1] Similarly, see Tuomela (2007, 135). While there are differences between Tuomela's and Gilbert's views, they will not make a difference to the argument developed here. One might worry that such an account is committed to voluntarism about belief. However, the individual beliefs on which the existence of the collective belief (in part) depends need not presuppose voluntarism, even if the collective beliefs to which they give rise are voluntarist. It is also worth noting that there is an important difference between the conditions for the creation of a belief and the conditions for the maintenance of that belief. There is more room for an exercise of will in the maintenance of a belief even in individuals than there is (on most accounts) in the acquisition of a belief.

[2] See e.g. Tuomela (2007), 136.

[3] Deborah Tollefsen is a nice example of someone who is attempting to do precisely this. She has criticized Gilbert's and Tuomela's accounts of group beliefs on the ground that 'our attributions of intentional states to organizations are often, if not always, made in ignorance of the intentional states of the members (even the operative members)' (2002, 396–7). One advantage of the communities and beliefs on which I shall focus is that we are not ignorant of the relevant intentional states of the group members. They have articulated their individual beliefs quite clearly at a number of stages in

And whether or not one can offer a satisfactory virtue-epistemic account of knowledge,[4] the notion of an epistemically virtuous collective is a topic that deserves further attention. Rather, these issues are set aside because there is an important set of questions regarding what is involved in the revision of group beliefs that will remain, no matter how these disputes about the best definition of knowledge are ultimately settled. Consequently, I shall focus on issues relating to group belief revision.

As Gilbert notes, collective beliefs can be remarkably difficult to change. To begin with, there is the basic challenge of changing the mind of not just one but many members of the group. How many members' minds must be changed is an interesting question. Gilbert herself has expressed reservations about ascribing mental states to a group if less than half of the group members experience it (2002, 133); whereas Raimo Tuomela has argued that there may be cases where a group believes something 'that possibly no single member finds privately acceptable (think of a case of voting in which no one's first choice is elected)' (2007, 137). While I prefer a criterion that would permit the ascription of a belief to a group even in (some) cases where fewer than half of the group's members believe it (in what Tuomela calls the I-mode; see Pettigrove and Parsons 2012), my central argument does not depend upon the outcome of this debate.

There is also the socio-epistemic challenge of changing people's minds in a context in which it is believed that many other people accept the belief that one is thinking about setting aside. We are profoundly influenced by the fact that other people believe some proposition, p (Cialdini 2001). Furthermore, we ought to be influenced by this fact (Zagzebski 2012, 52–74). So the belief that a number of other people believe p can give p a high level of credence within a population even when that population has been presented with strong evidence to the contrary.

Third, there is a distinctive set of normative issues raised by collective belief revision. By committing to a collective belief, an individual places herself under an obligation to the other members of the collective 'to constitute—as far as is possible—a body that believes that p' (Gilbert 2000, 41). This will include acting in ways that are consistent with the belief, such as asserting it in appropriate contexts and not calling it or beliefs that it necessarily presupposes or entails into question. There is some scope for an individual's personal view to diverge from the view of the group of which she is a part. Nevertheless, if she expresses her

---

the process of formulating the group's belief. So the argument I develop will not depend upon the success of Tollefsen's challenge.

[4] Jason Baehr (2011) mounts a compelling case against the attempt to define knowledge in virtue-epistemic terms.

disagreement it is incumbent on her to flag the fact that she is only speaking as an individual, rather than as a representative of the group. And when she does so, it raises questions regarding her fidelity to the group and she may risk losing her standing within it (Gilbert 2000, 41).

Finally, there is the structural challenge of getting a change of mind from the individual to the collective level. The structure of some collective agents may be such that this requires little more than that a significant number of those who make up the collective change their minds and are aware that others have too. Within other collectives, however, the change may require that key office-holders change their mind, or that the change comes to be reflected in authoritative documents that officially represent the mind of the group.

Gilbert offers the following as a way of characterizing the conditions that, if met, would constitute collective belief revision. 'For a body of which I am a member to change its beliefs requires something akin to an agreement to stop believing that p together and to start believing that q instead' (Gilbert 2000, 46). This characterization provides a useful starting point for the analysis of group belief revision. However, I shall argue that the account she has offered will need to be modified before it can adequately cover a range of cases of group belief revision that ought to be central to any account of group belief revision, but especially to Gilbert's account.

Gilbert builds her discussion of group belief revision around an imagined scientific community. I shall work with a different community that in many ways better exemplifies the conditions of group belief that she identifies, namely, the Presbyterian Church (USA). One reason for focusing on the PC(USA) is that, unlike most academic communities, which tend to have a rather loose structure and are fairly reticent about committing themselves to collective beliefs, the PC (USA) explicitly presents itself as a collective that believes a number of things. The current *Constitution of the Presbyterian Church (USA)* is composed of two parts: *The Book of Confessions* and *The Book of Order*. In relation to the first of these parts, the constitution states: 'In these statements the church declares to its members and to the world who and what it is, what it believes, and what it resolves to do. These statements identify the church as a community of people known by its convictions as well as its actions' (Presbyterian Church 2013, F-2.01). These confessional documents often begin with 'We believe…', and are sprinkled throughout with similar assertions such as 'We confess and acknowledge…' and 'We declare…'.

Furthermore, these collective beliefs have arisen out of a process that satisfies Gilbert's conditions for constituting a joint commitment. Presbyterians have a representative form of government that is built around regional units called presbyteries, which are composed of the clergy in a particular region plus lay representatives from each of the churches in that region. The standard process whereby something makes it into the constitution of the church is as follows. Initially a member of a presbytery presents a proposal to their presbytery for consideration. After the presbytery has discussed the proposal a vote is taken. If a majority in the presbytery favour the proposal, it is sent to the national governing unit, the General Assembly, for consideration.[5] The General Assembly, which is composed of clergy and lay representatives from around the United States, discusses the matter and then votes upon it. If a majority within the Assembly approve, then the proposal is sent back down to the presbyteries—this time all of the presbyteries—for their consideration. If a majority of presbyteries approve the proposal,[6] it is sent to the General Assembly once more for final ratification. By the time a proposal has made it through this process, the parties to the collective belief have expressed their 'willingness to be party to it in conditions of common knowledge' (Gilbert 2000, 40).

Multi-generational collectives pose a special challenge, insofar as one might be concerned about the ability of past members to commit current and future members to believe something. But here too, the Presbyterian Church (USA) is well situated to deal with the challenge. In the case of the confessions in the first part of the constitution, the beliefs that have been endorsed by the church are widely disseminated. Selections from the confessions are often incorporated within the liturgy of Sunday morning church services. They play an important role in the theological education of both laity and clergy, and when individuals within the church are set aside for leadership roles, they agree to be guided by them. Hence, if any group satisfies Gilbert's condition for constituting a body that believes that p, it is the Presbyterian Church (USA).

Not only does the PC(USA) satisfy Gilbert's conditions for collective belief, it also provides a useful test case for an account of collective belief revision. The church has long identified itself with the motto: *Ecclesia reformata, semper reformanda*—'the church reformed, always reforming' (Presbyterian Church 2013, F-2.02). This reforming spirit has expressed itself at various points in the church's history in, among other ways, the revision of parts of its constitution. By far the most dramatic of these revisions took place in 1967. At a structural level, there were two principal changes. The church went from having a single

---

[5] Occasionally the recommendation comes from a committee set up by the General Assembly and it skips the initial phase of being presented to a presbytery. But thereafter the process is the same.

[6] Some decisions require a two-thirds majority, others only a simple majority, i.e. one more than 50%.

confessional document—the *Westminster Confession of Faith* of 1646—to having a book of confessions containing historic documents from multiple time periods and locations. The second structural change was the addition of a newly written confession, the Confession of 1967.[7]

These structural changes brought with them important changes in content as well. And some of the changes look as though they fit Gilbert's account quite nicely. For example, prior to 1967 leaders within the Presbyterian Church were asked to affirm 'that the Westminster standards contained the system of doctrine taught in Scripture'.[8] The standards in question were those produced by a committee set up by the English Parliament that met in Westminster Abbey from 1643 to 1646, which were adopted by the Church of Scotland in 1647 and subsequently endorsed by the English and Scottish parliaments. As one might expect, the moral and political worldview of American Presbyterians in the 1960s was somewhat different than the worldview of English and Scottish theologians and parliamentarians at the time of the English Civil War. The Westminster Divines, for example, insisted that it was the duty of the civil magistrate to suppress 'all Blasphemies and Heresies' and prevent 'all corruptions and abuses in Worship and Discipline' (Leith 1982, 220), and it is clear that they considered the Roman Catholic Church to be among those whose teachings should be suppressed.[9] Well before the 1960s American Presbyterians had removed references to the suppressing of heresies, and so on, from the magistrate's list of duties and had inserted in its stead, 'no law of any commonwealth should interfere with, let, or hinder, the due exercise thereof, among the voluntary members of any denomination of *Christians*, according to their own profession and belief' (Presbyterian Church 1999, 6.129). The Confession of 1967 takes the matter a step further, asserting that: 'The Christian finds parallels between other religions and his own and must approach all religions with openness and respect' (Presbyterian Church 1999, 9.42).

If one were to represent this shift in Gilbert's terms, one might use p_1 and p_2 to represent the propositions, 'Christians should reject the teachings of Catholics, Hindus, Muslims, etc. as false' and 'The civil magistrate should suppress them' and q_1 and q_2 to represent the propositions, 'The civil magistrate should not suppress the teachings of Catholics, Hindus, Muslims, etc.' and 'Christians should be open to and respectful of the truths they might disclose'. One might

---

[7] Although it was the most dramatic, it was also one of the most widely supported changes, with nearly 90% of presbyteries endorsing the Confession of 1967 (Loetscher 1983, 162).

[8] Willis (1988), 119. For other changes made to the 'subscription formula' in 1967, see Loetscher (1983), 164–5.

[9] For example, the Pope is referred to as the Antichrist (Leith 1982, 222).

then say that at some point between 1647 and 1967 the PC(USA) agreed to stop believing together that p_1 and p_2 and to start believing that q_1 and q_2 instead. However, even if Gilbert's account of collective belief revision works for some cases, it does not work for others. And even some of the cases it appears to fit can be more perspicuously described. Or so I shall argue. However, before I present that argument it will be useful to get some other conceptual tools in place first.

2. From Epistemology to Collective Moral Epistemology

There are a number of different things that have marched under the banner of 'knowledge'. One of these was mentioned above, according to which an agent with knowledge of p, (a) believes that p, (b) is warranted in believing that p, and (c) p is true.[10] Call this *propositional knowledge-that*. Jonathan Kvanvig (1992) and others have criticized formulations like this, arguing that while they fit some instances of knowledge they fail to capture others. One problem, Kvanvig argues, is that they are too focused on propositions, which are treated as discrete atomic units that are isomorphic with the sentences of a natural language. Many instances of knowledge do not have such propositions as their objects. He gives as an example the knowledge someone has after standing on the edge of the Grand Canyon and gazing at it for a few minutes. If asked afterward whether the canyon is longer than it is wide or deeper than an elephant is tall, the observer would confidently answer, 'Yes'. But this is not because those two propositions were among the bundle of discrete atoms of knowledge that he acquired while standing at the canyon's edge and which he is now carrying around in his head. The object of knowledge that he acquired was something more holistic, which he is able to carve up in various ways when he finds it useful to do so (Kvanvig 1992, 180–2). For convenience let us call this *holistic knowledge-that*.

In recent years a number of people have suggested that, in addition to the kinds of knowledge with which contemporary epistemologists have primarily been concerned, there are also other truth-related states that at earlier points in the history of western philosophy were the main concern of epistemologists. Knowledge so construed, which Linda Zagzebski calls 'understanding', is a matter of seeing or grasping relations between things, concepts, meanings, or propositions.

---

[10] An agent is warranted in believing a proposition p when an agent's getting to the truth about p isn't merely the result of certain kinds of luck. This description is not meant to take a side in current debates about what constitutes knowledge. The aim here is simply to get on the table a range of things that might go by the name of knowledge.

It is exemplified by the scientist who not only possesses knowledge of discrete facts but grasps the relationships that hold between them, fitting them into an overarching theoretical framework, seeing them as parts of a larger whole. It is manifested by the skilled reader who does not merely follow the basic plot of a novel, for example, but recognizes themes that run through the work, or through the author's wider corpus, or through the western literary tradition, or that connect the work with other events in the world as she interprets the novel (Zagzebski 2001, 237; Baehr 2011, 146–8). And it is disclosed in the philosopher's recognition of relations of entailment, necessity, and possibility and the mathematician's comprehension of a proof.[11]

Each of the kinds of knowledge mentioned above can be seen in contexts where the object of concern is moral knowledge. An agent can have knowledge-that of moral propositions as well as holistic moral knowledge-that. She can also have moral understanding, grasping moral concepts, meanings, and relations that might hold between them. 'Moral understanding', Zagzebski observes, 'includes seeing the connection between moral reasons and moral judgments, and perhaps also the connection between certain emotions and moral judgments. Understanding permits us to see how to extend a moral judgment to different situations, and to see how distinct moral judgments relate to each other' (2012, 175). However, even if one prefers to work with a narrower conception of knowledge that excludes epistemic states like understanding, Allan Gibbard has suggested that one need not be troubled by talk of moral knowledge. There may be differences between moral knowledge and knowledge of other kinds, but whatever their differences, it is reasonable to think their rough outlines will be similar enough that we can treat them as essentially of a piece.[12]

To this point the examples we have discussed have been examples of things known by individuals. However, it would seem that we can draw the same

---

[11] Zagzebski (2012), 175. Stephen Grimm (2006, 517–18) has objected that what Zagzebski calls understanding is not as different from the kind of knowledge with which contemporary epistemologists have been concerned as she thinks. However, his argument is built around confusing what Miranda Fricker (1998, 164–6) calls 'the philosopher's question' with what she calls 'the historian's question'. Grimm starts with the claim that 'our understanding of natural phenomena' is 'arguably the paradigm case of understanding'. And the illustration that follows shows that what he has in mind is identifying the causal explanation of a particular natural occurrence (the historian's question). However, Zagzebski is concerned with a much more diverse range of paradigmatic cases: music, art, literature, mathematical proofs, conceptual relations, metaphysical relations, and philosophical and scientific theories. And these other cases do not share important features with the understanding of natural phenomena on which Grimm's argument depends. Consequently, Zagzebski's account of understanding is not (or not obviously) subject to Grimm's objection.

[12] He argues that both the moral realist and the expressivist can accept such a claim (Gibbard 2002, 224–6).

distinctions with respect to collective knowledge that we do with respect to individual knowledge. We can refer to collective propositional knowledge-that, collective holistic knowledge-that, and collective understanding, as well as to collective moral knowledge of each of these types. One might worry that the ascription of knowledge to collectives requires one to buy contentious metaphysical assumptions about collective mental states. However, this does not follow from references to collective knowledge or collective beliefs any more than it follows from a theorist's use of folk concepts like belief, thought, and desire when talking about individuals that she is committed to a metaphysical framework in which these are distinct pieces of mental furniture that possess causal powers similar to the powers of material objects. We can ascribe collective knowledge of the above-mentioned types without taking a stand on whether collective knowledge and collective belief are (a) novel emergent properties that supervene on properties of the individuals who make up the collective, or (b) a shorthand way of referring to qualities that are ultimately reducible to properties of individuals.

Consequently, the move from a discussion of individual knowledge to a discussion of collective knowledge is less metaphysically contentious than it might seem. And making such a move can draw our attention to epistemic qualities of groups that might otherwise go unnoticed. One such quality is the surprising reliability of the average judgment of certain sorts of groups on a wide range of questions. Francis Galton (1907) drew attention to group reliability in the early twentieth century when he noticed that the statistical average of the judgments of people at an English fair who estimated the weight of an ox came within 1 per cent of the actual weight of the animal. This result had been anticipated in the eighteenth century by Nicolas de Condorcet's jury theorem, and similar results have been replicated in cases involving estimating the temperature in a classroom, producing a rank-ordering of the sizes of very similar piles of buckshot, predicting the winners of baseball playoffs and American presidential elections, and even in choosing the best move in a game of chess (Sunstein 2006, 21–43).

Another domain that comes more clearly into view when we attend to the epistemic attributes of collectives is the body of knowledge that is commonly referred to as tradition. A tradition is a way of making sense of ourselves and our world that is shared with and inherited from others: 'What we receive from the past are, in effect, beliefs, persuasions, convictions; that is, ways of "holding for true"...' (Ricoeur 1988, 222–3), 'ways of seeing and of speaking about the world', ways of acting and interacting (Williams 1983, 12, 20–1; MacIntyre 1983, 221–2). Our tradition reflects the wisdom (and sometimes the folly) of our forebears and shapes our sense of who we are, who we aspire to become, what we are doing, and what we ought to be doing. It is preserved in rituals, texts, habits, social structures, architectural styles, and countless other forms and features of life.[13] Born as we are into a tradition-saturated world, we imbibe it from our earliest moments, as we learn to think, speak, and act. 'We are carried along by it before we are in a position of judging it, or of condemning it' (Ricoeur (1988, 223). It makes and makes-sense-of our characteristic ways of going on, and it is revealed both in what we think about self-consciously and in what we take for granted when thinking (Gadamer 1989, 276–7; Taylor 2004, 23–30).

3. Traditions and Collective Belief Revision

With these distinctions in place, we are now in a better position to see why Gilbert's account of collective belief revision is inadequate as it stands. The first reason is its focus on belief as a state whose object is a proposition.[14] If we can have holistic knowledge-that and if the holistic qualities of our epistemic experiences can be shared by a group of which we are a part, then an account of belief revision that characterizes the object of collective belief as a proposition will be incomplete, insofar as it leaves out an entire category of beliefs, namely those involved in collective holistic knowledge-that.[15]

There is good reason to think that if individual believers have such non-propositional belief contents, collective believers could, too. To revert to Kvanvig's Grand Canyon example, one could imagine a group of tourists standing at the edge of the canyon forming a collective belief on the basis of their shared experience. If one requires joint commitment as a condition for collective belief, then to some degree their shared experience may need to be put into words. But these words might themselves simply be a placeholder for the not yet articulated—and perhaps not fully articulable—non-propositional belief content.

---

[13] Compare what we have described as a 'tradition' with what Tuomela refers to as a group's 'ethos: groupish 'thinking and acting involves the existence of an ethos, namely, some constitutive goals, values, standards, and norms and we-mode thinking and acting relative to them' (2007, 146). In most cases a group's ethos will consist in a subset of the goals, values, standards, and norms that (partly) make up the group's tradition.

[14] Tuomela likewise focuses his discussion on propositional knowledge-that and on beliefs that take propositions as their objects (see 2007, 135–6).

[15] Some readers might prefer to account for the phenomenon Kvanvig highlights by appealing to a dispositional account of belief, rather than positing beliefs with holistic content. A dispositional account is quite attractive for those who wish to draw attention to collective beliefs, since collective dispositions might seem less metaphysically spooky than collective mental states. Nevertheless, dispositional accounts of belief still face a number of unanswered questions. So at present it is not clear than a dispositional account is preferable to the holistic one offered above.

For example, one of them might say, 'The Grand Canyon is big', and the others might agree to share this belief, without any of them thinking that the proposition 'The Grand Canyon is big' captures the full content of their shared belief. Indeed, the expression of shared belief need not wear the guise of a proposition at all. Standing at the canyon's edge, one of them might simply say, 'Wow!' And the others might echo, 'Yeah, Wow!' Against the backdrop of a standing commitment within the group to share beliefs based on shared experiences,[16] these expressions of wonder could be sufficient to give rise to a shared belief with holistic content. Within the context of a religious community that gathers regularly for prayer, meditation, and worship, whose members share liturgically mediated religious experiences, and who are jointly committed to sharing religious beliefs, we should expect shared beliefs with holistic content to be fairly common.

Hilary Putnam draws attention to another reason we might expect such communities to share beliefs with holistic content. Often practical reasoning involves imagining what it would be like to pursue the different courses of action that are available to us at a particular moment. The mountain climber who is considering two different possible routes to the top imagines what each route will involve and uses this process to direct his next step.[17] The quandary the imaginer faces may also run deeper than this, perhaps calling into question who she is or who she should become. The person in mid-career who is considering a change of vocation will imagine the alternative career path and compare it with what she imagines her future will look like if she remains in her current vocation.[18] Our imaginative engagement with these possibilities can enable us to see which course we should take. Putnam observes:

> this sort of reasoning need not at all be reducible to any kind of linear proposition-by-proposition reasoning... Of course, saying that this is not linear propositional reasoning is not to deny that *after* he has imagined 'what would happen if...' he can *put* the relevant considerations into words. It is to say that he need not have *stored* the relevant information *in the form of words*. (Putnam 1978, 86)

---

[16] Oliver O'Donovan claims that such a commitment to being in agreement is definitive of the church (2008, 30–4).

[17] Putnam (1978), 85–7. It was precisely the need to make room for such cases and the knowledge they might bring that prompted Christopher Peacocke (2003) to revise his account of concepts by introducing the idea of 'implicit conceptions'.

[18] As Charles Taylor points out, these imaginative engagements may involve quite a bit more than simply weighing future goods that one expects will accompany each of these career paths. The options may also involve competing conceptions of who one is, what the decision amounts to, and what really matters (1985, 26–7).

Preaching often makes use of precisely this sort of imaginative engagement, as is illustrated in Frederick Buechner's classic sermon, 'A Sprig of Hope':

> Let us start with the story itself; more particularly, let us start with the moment when God first spoke to Noah; more particularly, let us start with Noah's face at that moment when God first spoke to him. When somebody speaks to you, you turn your face to look in the direction that the voice comes from; but if the voice comes from no direction at all, or the voice comes from within and comes wordlessly, and more powerfully for being wordless, then in a sense you stop looking at anything at all... Your face goes vacant because for the moment you have vacated it and are living somewhere beneath your face, wherever it is that the voice comes from. So it was maybe with Noah's face when he heard the words that he heard, or when he heard what he heard translated clumsily into words: that the earth was corrupt in God's sight, filled with violence and pain and unlove—that the earth was doomed.
>
> It was presumably nothing that Noah had not known already, nothing that any man who has ever lived on this earth with his eyes open has not known. But because it came upon him, sudden and strong, he had to face it more squarely than other people usually do, and it rose up in him like a pain in his own belly. And then maybe, like Kierkegaard's Abraham, Noah asked whether it was God who was speaking or only the pain in his belly; whether it was a vision of the glory of the world as it first emerged from the hand of the Creator that led him to the knowledge of how far the world had fallen, or whether it was just his pathetic human longing for a glory that had never been and would never be. If that was his question, perhaps a flicker of bewilderment passed across his vacant face—the lines between his eyes deepening, his mouth going loose, a little stupid. A penny for your thoughts, old Noah.
>
> But then came the crux of the thing, because the voice that was either God's voice or an undigested matzoh ball shifted from the indicative of doom to the imperative of command, and it told him that although the world was doomed, he Noah, had a commission to perform that would have much to do with the saving of the world.
>
> (Buechner 1994, 228–9)

The preacher invites the gathered community into a story, often a story that plays an important role in structuring and preserving this community's sense of who it is and how it ought to live, a tradition-shaping story. Engagement with the story enables members of the community to reflect on their own past experiences and to anticipate a range of responses to future experiences.[19] It encourages them to notice certain features of their circumstances that might otherwise have gone unnoticed and offers them alternative ways of making sense out of those and

---

[19] Oliver O'Donovan claims: 'We focus our attention on the good presented to us by approaching it through narrative and projecting it through hope' (2008, 99). He argues that this is a distinctive feature of being a temporally extended agent that experiences the present 'as set "between" past and future'. Nevertheless, he suggests, Christian theology provides this common activity with a distinctive flavour and rationale for members of the Christian community.

other aspects of their lives. This sort of imaginative engagement through story can function like the shared experiences of the group standing at the edge of the Grand Canyon. It can provide the object of their joint belief. And when the story invites them to reflect on future decisions the group might make—as Martin Luther King, Jr.'s 'I Have a Dream' address did for many churches—the group's imaginative engagement might function like that of the mountain climber deliberating between two routes or the person in mid-career deliberating about vocational paths.

To accommodate these sorts of cases, we might modify Gilbert's account by adding variables ($p+$ and $q+$) to represent the sorts of objects that beliefs take in cases of holistic knowledge-that:

A group g revises its belief that p or $p+$ iff the members of g (or the operative members of g) agree to stop believing that p or $p+$ and start believing that q or $q+$.[20]

Such an account will be better suited to manage a wider range of cases. However, it does not yet leave room for the kind of belief revision involved in acquiring a greater depth of understanding. To begin with a familiar example, consider what is involved in coming to understand *modus ponens*. A student might believe p, and $p \rightarrow q$ (indeed she might even believe q), but fail to see that q follows from the conjunction of the other two propositions. The student who finally comes to understand *modus ponens* has a different belief set than she did beforehand. But she has neither rejected a belief she previously held nor simply added another proposition to her prior set of beliefs (Carroll 1895). Rather, she has come to understand something new about the relationship between the objects of beliefs she already accepted. And if her prior beliefs that p, $p \rightarrow q$, and q were warranted and true, we might say she has come to understand what she already knew.

A similar sort of drawing together and making new connections between distinct and perhaps already familiar beliefs can be seen in the context of moral

---

[20] This is the first in a sequence of proposed definitions of group belief revision. Two features of the definitions in this sequence are worth noting at the outset. First, the sequence illustrates what it is describing, namely, a process of belief revision. Second, it departs from the mainstream of logicians working on belief revision who take their bearings from Alchourrón, Gärdenfors, and Makinson (1985) and Gärdenfors (2008). A trivial departure is that I am using 'belief revision' to cover both the kind of change generated by attempting to resolve a contradiction in one's belief set (which they call 'belief revision) and the kind of change involved in adding a new belief to an existing set of beliefs (which they call 'belief expansion'). The more important departure is that these logicians have been attempting to model belief revision using propositional logic. However, it is unclear that a move from $p+$ to $q+$ can be modelled perspicuously using propositional logic. The same is true for the changes in conception discussed below. Modelling these kinds of changes will require the use of at least predicate logic. It is hoped that these definitions will identify the kinds of factors logicians will need to accommodate when developing more nuanced models of belief revision.

and theological understanding, as well. For example, the Confession of 1967 used the theme of reconciliation—borrowed from a passage in Paul's second letter to the Corinthians—to reorganize and reinterpret a number of the church's historic doctrines in a way that highlighted their relevance to social issues at the forefront of American consciousness in the 1960s. The result was a call to social action that was rooted in a distinctive understanding of who God is, who the church is, and what the church is called to do that differed in a number of respects from the vision articulated in the Westminster Confession. The Westminster Divines did not feel the need to address questions of racial justice, and for more than two centuries many of those churches who used some version of the Westminster Confession as their statement of faith deemed it compatible with a system whereby members of one race enslaved those of another.[21] By contrast, the Confession of 1967 asserts, 'the church labors for the abolition of all racial discrimination and ministers to those injured by it. Congregations, individuals, or groups of Christians who exclude, dominate, or patronize their fellowmen, however subtly, resist the Spirit of God and bring contempt on the faith which they profess' (Presbyterian Church 1999, 9.44). Similarly, even though they are drawing on shared theological resources, the Westminster Divines would have been surprised by the critique of nationalism expressed in the Confession of 1967, as well as by the claim that: 'The church, in its own life, is called to practice the forgiveness of enemies and to commend to the nations as practical politics the search for cooperation and peace' (Presbyterian Church 1999, 9.45). To accommodate the possibility of such revisions in collective belief, we need to add what we might call an *understanding operator* (u) to our account and remove the necessity of ceasing to believe that p. If we let u(Rpq) stand for an understanding of a relationship that exists between p and q, then:

A group g revises its belief that p or $p+$ iff the members of g (or the operative members of g), who previously believed that p or $p+$, agree to believe something new, such as (i) that not-p, or (ii) that q or $q+$, or (iii) that p or $p+$ and u(Rpq).

Revisions of belief that involve changes in moral understanding characteristically involve another sort of change that it would be useful to capture in our account of collective belief revision, namely, changes in conception. Our most primitive evaluative judgments are built around encounters with exemplars (Zagzebski 2004, 40–50). Some of these exemplars are ordinary people who happen to be

---

[21] In addition to being endorsed by Presbyterians and (briefly) Anglicans, the Westminster Confession 'was adopted with modifications by Congregationalists in England and New England, and it was the basis of the Baptist creeds, the London Confession, 1677, 1688, and the Philadelphia Confession of Faith, 1742' (Leith 1982, 192–3).

among our narrow circle of associates at an early stage in our lives. Others may stand out in ways that come to be recognized in a much wider community. Let us imagine some such individual who has been rightly picked out by the community as an especially admirable person. In referring to what is admirable about this person, the members of the community use some word—for example, just or loving—to designate the trait in question. They may also discuss the nature of the property they have picked out. Since their ability to pick out the exemplar differs from their ability to describe what is admirable about her, it would not be surprising if this discussion were quite illuminating in some respects, even though it was obfuscating in others. Subsequent speakers, perhaps even subsequent generations of speakers, may be left to work out a more faithful or complete account of the nature of the quality first identified by the community. This process may be complicated by the way in which earlier speakers attempted to describe the nature of the quality, since some of what they built into the description may mislead. Furthermore, given the way our concepts shape the world we see, once the concepts we are working with have been influenced by prior descriptions, we are likely to see the exemplar in a distinctive way that reflects details of these descriptions.

Something like this story seems plausible as an account of the origins of many of the moral judgments preserved within a tradition. And as W. B. Gallie noted in his seminal paper on the topic, it is also an ideal recipe for the development of essentially contested concepts, that is, 'concepts the proper use of which inevitably involves endless disputes about their proper uses on the part of their users' (1955-6, 169). When combined, (a) the complex nature of the quality in question, (b) the equally complex range of the contexts in which it might or might not be appropriately manifested, and (c) the different vantages from which observers might perceive and make sense of the quality, can easily generate competing specifications of the quality in question—what John Rawls called rival conceptions of the concept or quality.[22]

Gallie himself contended that many of our ethical concepts are essentially contested. Whether one thinks this is correct as a claim about secular ethics or not, it is exceedingly plausible in the case of theological ethics. Theological ethics is concerned with how one should live in the light of what one takes to be a divine reality. Christian theological ethics, which is the variant most relevant to the example we have been discussing, is concerned with how one should live 'in the

---

[22] Rawls (1999), 427. For further discussion of the nature of essentially contested concepts and the challenges that might be faced by the claim that justice, for example, is such a concept, see Christine Swanton (1985).

presence of a gracious God' (Tracy 1996, 52), whose nature was revealed in the person, work, and teachings of Jesus of Nazareth (Presbyterian Church 1999, 9.07–15). Thus, Christian theological ethics combines the features of the historic exemplar story above with 'a mystery beyond the reach of man's mind' (Presbyterian Church 1999, 9.15). This means that communities like the PC(USA) will find themselves continually drawn back to important accounts of what early Christians heard and saw of Jesus and his ministry. But at the same time, there are at least three reasons why they should find it challenging to understand what was heard and seen. First, the initial observers had difficulty understanding the person they saw and heard.[23] Second, the way in which the twenty-first century community interprets those early texts is mediated by two thousand years of interpretation and ritual which themselves originated in and were shaped by remarkably different cultural and conceptual horizons. Third, the reality being disclosed, in the light of which the community seeks to live, is by its very nature supposed to be beyond the bounds of the community's comprehension.[24] Consequently, we should expect the moral reflections of communities like the PC(USA) to be marked by an interesting tension between a conservative and an innovative impulse, between recovering and discovering, remembering and revising.[25]

One of the more significant changes expressed in the Confession of 1967 concerned precisely this tension. It involved rival conceptions of the inspiration and authority of scripture. The conception advanced by the Westminster Confession claimed that the books of the (Protestant) Old and New Testaments were 'immediately inspired by God, and by his singular care and providence kept pure in all ages' (Presbyterian Church 1999, 6.008). Under the influence of several generations of Princeton theologians in the nineteenth century, this conception was taken to entail, among other things, 'the inerrancy of the Bible'. And starting

---

[23] One of the more endearing literary features of the gospel narratives is the way in which they convey the disciples' deep and persisting confusion regarding who Jesus was and what he was doing. Far from portraying the first generation of Jesus-followers as having all the answers, the gospels show them still not getting it, even after three years of intense tutoring.

[24] In a similar vein, John Cottingham observes, 'it is the task of religious discourse to strain at the limits of the sayable' (2003, 8).

[25] Rowan Williams, the recently retired Archbishop of Canterbury, explores the importance of this tension in a number of his publications. For example, in his work on Arius he argues: 'The doctrinal debate of the fourth century is thus in considerable measure about how the Church is to become intellectually self-aware and to move from a "theology of repetition" to something more exploratory and constructive' (1987, 235); and in 'What Is Catholic Orthodoxy?' he argues that one of the distinguishing features of orthodoxy is that, in the very act of seeking to be faithful to its source it also 'always subverts its own finality—as a system of words or pictures' (1983, 19). For a nice overview of Williams's work on this theme, see Benjamin Myers (2009).

in 1910, endorsement of this inerrantist conception was made compulsory for all candidates for ordination (Rogers 1991, 205). The Confession of 1967's comments on Scripture, by contrast, are rooted in an alternative conception of inspiration and authority.

The Scriptures, given under the guidance of the Holy Spirit, are nevertheless the words of men, conditioned by the language, thought forms, and literary fashions of the places and times at which they were written. They reflect views of life, history, and the cosmos which were then current. The church, therefore, has an obligation to approach the Scriptures with literary and historical understanding.   (Presbyterian Church 1999, 9.29)

According to this latter conception, the inspiration and authority of Scripture was compatible with it also being a human—sometimes all too human—document that reflected the human author's imperfect, albeit instructive, understanding of divine reality. And this conception carried with it new possibilities for thinking about what it means for the actions of today's PC(USA) to be guided by these ancient texts.

This alternative conception of authority was further supported by the expansion of the Presbyterian constitution to include a larger number of confessional documents. For a considerable portion of its history, candidates for ministry in the church had to indicate their agreement with the Westminster Confession. If they disagreed with the Westminster Divines on any point, they had to identify that fact and defend their position before the Presbytery. However, the expansion of the 'subordinate standards' contained in the Book of Confessions in 1967 reflected a decisive shift in perspective. Some of the documents that were added, like the Nicene Creed and the Apostles' Creed, were already routinely used in Presbyterian liturgy. Others, like the Scots Confession and the Heidelberg Catechism, had played an important role in the church's past and continued to play an important role in other contemporary denominations in the Reformed tradition, but had fallen out of use in most Presbyterian Churches. Still others, in particular the Declaration of Barmen and the Confession of 1967, were twentieth-century documents explicitly written as a response to the needs of their current social and political moment. The most interesting change introduced by this expansion of the church's constitution arose out of the multiplicity of voices that were brought to stand alongside those of the Westminster Divines. This represented a shift in thinking that had already taken place in the church, which the new structure was instituted to reflect as much as to create. It represented a different conception of theology as well as of the kind of guidance offered by confessional documents. The inclusion of multiple documents from different places and times encouraged a greater diversity of theological perspectives. It also drew attention to the contextual nature of theology, inviting church members to think of theology not so much as a timeless statement of eternal truths to be memorized and repeated but rather as a tool for engaging with particular questions that arise in particular situations. At one point in the church's history, these were genuine innovations. But the most natural way to describe what was taking place in 1967 is not in terms of the rejection of one belief and the endorsement of another. Rather, what happened was that the institutional structure was modified to reflect an underlying set of beliefs that were already in place.

Given the nature of moral communities like the PC(USA), we should expect a number of instances of belief revision to involve a shift from one way of conceptualizing some feature of who they are and who they are meant to be to another. In many cases the concepts will remain fixed but the conceptions will change. To accommodate such changes, it will be useful to introduce another feature to our account to represent conceptions. Let c_1 and c_2 stand for rival conceptions. We can then modify our account of collective belief revision as follows:

A group g revises its belief that p or $p+$ iff the members of g (or the operative members of g), who previously believed that p or $p+$, one of whose concepts is construed in the c_1 way, agree to believe something new, such as (i) that not-p, or (ii) that q or $q+$, or (iii) that p or $p+$ and u(Rpq), or (iv) that p or $p+$, one of whose concepts is construed in the c_2 way.

This account highlights a number of aspects of collective belief revision that Gilbert's original account either lumped together or excluded entirely. However, for all we have said, there may be other instances of collective belief revision that our modified account fails to accommodate. The easiest way to leave room for this possibility is to replace the biconditional in the preceding variant with a simple conditional. This would yield:

A group g revises its belief that p or $p+$ if the members of g (or the operative members of g), who previously believed that p or $p+$, one of whose concepts is construed in the c_1 way, agree to believe something new, such as (i) that not-p, or (ii) that q or $q+$, or (iii) that p or $p+$ and u(Rpq), or (iv) that p or $p+$, one of whose concepts is construed in the c_2 way.

Conclusion

Contemporary discussions of collective actions, emotions, beliefs, and knowledge have benefited greatly from Margaret Gilbert's work on these topics. Her discussion of group belief revision, likewise, has drawn attention to an important but under-theorized aspect of collective life. However, perhaps because it focused on an imagined scientific community, her discussion failed to draw attention to a number of features of belief revision that play a significant role in the lives of many collectives. By contrast, the constitutional changes made by the Presbyterian Church during the middle part of the last century provide a nice illustration of revisions of group belief that are not readily covered by Gilbert's account. For the most part the lack of fit is due to the fact that the revisions were not (or not simply) a matter of agreeing to reject one proposition and replace it with another. Instead, they involved changes in conception, understanding, or holistic knowledge.

As should already be clear from the preceding discussion, the need for a revised account is not limited to religious communities like the PC(USA). It will apply to a wide range of communities, but especially to those whose identities are built around normative commitments. Julia Annas has observed that 'each generation alters its predecessor's conception of some virtues, while others fall out of favour altogether' (2011, 22).[26] These generational shifts will affect community groups and charitable organizations like Rotary Clubs and Oxfam. It will be relevant to the beliefs and decisions of school or university trustees whose actions and membership are shaped around a founding trust deed. And it will have a bearing on a number of more loosely structured moral communities. To account for the kinds of revisions that take place in these moral communities will require a more diverse account of group belief revision than the one Gilbert provided. The account developed in section 3 goes a considerable way toward addressing this need. But it is only the beginning of what will need to be a much longer conversation. Sooner or later it, too, will need to be revised.[27]

---

[26] Similarly, MacIntyre suggests: '[A]ll reasoning takes place within the context of some traditional mode of thought, transcending through criticism and invention the limitations of what had hitherto been reasoned in that tradition...A living tradition...is an historically extended, socially embodied argument, and an argument precisely in part about the goods which constitute that tradition' (1983, 222).

[27] This chapter began in a conversation with Dan Speak and I am grateful for the many hours he spent discussing the topic and commenting on multiple drafts. The chapter has also benefited from feedback provided by Garrett Cullity, Antony Eagle, Patrick Girard, and Rod Girle.

PART III
Political Philosophy

7

The Epistemic Circumstances of Democracy

Fabienne Peter

Introduction

Does political decision-making require experts or can a democracy be trusted to make correct decisions? This question has a long-standing tradition in political philosophy, going back at least to Plato's *Republic*. Critics of democracy tend to argue that democracy cannot be trusted in this way, while its advocates tend to argue that it can. While they come to different conclusion about the epistemic value of democracy, both camps share an epistemic instrumentalist conception of democratic legitimacy and of political legitimacy more generally. By epistemic instrumentalism I mean the view that epistemic value derives from epistemic outcomes. Applied to democratic legitimacy, the view is that the legitimacy of democracy depends on its instrumental epistemic value. On this view, if there is a correct decision—an outcome that is truly just or truly for the common good, or at least more just or closer to the common good than others—then the legitimacy of democracy depends on how well suited it is to track this decision.

In contemporary political philosophy many epistemic democrats have embraced this epistemic instrumentalist defence of democracy and argued that democracy is a good means—or at least a good enough means—to reach correct decisions. But here is a challenge for this view, well articulated, I find, by Hans Kelsen (1955, 2):

[T]he doctrine that democracy presupposes the belief that there exists an objectively ascertainable common good and that people are able to know it and therefore to make it the content of their will is erroneous. If it were correct, democracy would not be possible.

The challenge, as I interpret it, is that an epistemic instrumentalist defence of democracy is self-undermining because the epistemic circumstances it presupposes are incompatible with democracy. Specifically, as I shall explain, the problem arises from what the epistemic instrumentalist conception of democratic legitimacy presupposes about epistemic authority.

To illustrate the challenge, consider a situation in which there seems to be no epistemic case for democracy. Suppose a town is considering the plan to build a new bridge across the river that runs through it. The decision on whether or not to build the bridge depends only on one factor, namely on the stability of the planned bridge. And suppose the town engineer has the expertise to assess whether the planned bridge is stable and concludes that it is.

In cases such as the bridge case, the verdict of the town engineer appears to be sufficient to legitimize the decision that the bridge should be built. It would be redundant, or perhaps even crazy, to seek a democratic decision on whether or not the bridge is stable. The fact that the town engineer enjoys epistemic authority over the matter thus undermines the epistemic case for a democratic decision on this issue.

The challenge, as I interpret it, is this. If there is a correct decision to be made and if someone has legitimate epistemic authority to make claims about what the correct decision is, the epistemic case for democracy crumbles. A first aim of my chapter is to show how the epistemic instrumentalist attempt to make democratic legitimacy conditional on the epistemic quality of the outcomes of democratic decision-making runs into a version of this challenge and should be rejected for that reason.

What are the alternatives to an epistemic instrumentalist defence of democracy? The first is to abandon the epistemic strategy altogether and defend democracy on practical grounds. This way of responding to the challenge leads to "deep proceduralist" (Estlund 2008) conceptions of democratic legitimacy, according to which democracy is legitimate not because it tracks a procedure-independent truth, but because the decision-making-procedure embodies (moral) values such as equality, dignity, and so on which confer value to its outcome. This is not the strategy I shall pursue here.

The second alternative is an epistemic proceduralist approach. It preserves a central role for epistemic considerations in the justification of democracy. But it brings such considerations to bear on the evaluation of the democratic decision-making procedure directly, not indirectly via the outcomes it produces. This is the strategy I shall explore in this chapter.

I shall use epistemic considerations to say something about the appropriate scope for democratic decision-making. Questions about the appropriate scope for democratic decision-making are typically asked in socio-spatial terms: who should be included in the democratic collective?[1] But it seems to me that we should also ask which issues should be subjected to democratic decision-making. This question, I want to argue, is in the first place an epistemic question: under what specific epistemic circumstances is democratic decision-making—as opposed to, most importantly, decision-making by experts—appropriate and thus potentially legitimate?

In cases like the bridge case, these circumstances are not given; democratic decision-making will most likely be illegitimate in this case. Critics of democratic decision-making have a point when they identify cases like the bridge case to argue for the unattractiveness of democracy. But not all cases are like the bridge case, and this leaves room for advocates of democracy to defend their cause. I will argue that when there is no procedure-independent epistemic authority about what the correct decision is, then there is a prima facie epistemic case for democratic decision-making.

The title of my chapter, "the epistemic circumstances of democracy", borrows from David Hume's and John Rawls' idea of the "circumstances of justice" and Jeremy Waldron's idea of the "circumstances of politics". Hume (1978 [1739]) and Rawls (1971) argue that justice has its natural place in circumstances of moderate scarcity and limited altruism. Waldron (1999) argues that politics has its natural place in circumstances in which there is a need for collective action but where people disagree about what to do. I accept this characterization of the circumstances of politics. But Waldron hasn't explained why disagreements need to be taken seriously. I shall make use of the epistemology of disagreement to help identify the appropriate locus of democratic decision-making.

My main focus shall be on deliberative democracy. Let me explain briefly what I mean by this term. Deliberative democracy is usually understood in contrast to aggregative democracy. Aggregative theories of democracy take the key feature of democracy to be the aggregation of individual preferences or beliefs through voting, where each person's vote is given equal weight. Theories of deliberative democracy, by contrast, view democratic decision-making as embedded in an exchange of reasons for preferring certain outcomes or for believing certain facts. They take the deliberation among the members of the democratic collective, again under some conditions of equality, to be an important justifying feature of democracy. Such public deliberation may take place formally, for example, in parliament, in the media, in meetings and events of political parties and other

---

[1] See the literature on what is called the "boundary problem" or the "constitution of the demos" (e.g. Goodin 2007; Miller 2009).

political organizations, and, informally, in people's discussions with their friends, colleagues, and family members. Since such deliberation is unlikely to produce a consensus, however, even a deliberative theory will assign some role to aggregative decision-making. A stark contrast between the two theories of democracy thus overstates the case. As I see it, the main difference between the two theories is that deliberative democracy does whereas aggregative democracy does not assign a legitimizing role to public deliberation. It is in this sense that I shall refer to deliberative democracy.[2]

1. Practical and Epistemic Authority

My argument against epistemic instrumentalism hinges on the relationship between epistemic authority and the legitimate practical authority of democracy. Before I can get into the argument, I need to say something about what I mean by these terms.

By legitimate practical authority I mean here the right to make claims which give others sufficient reason for action. When a legitimate practical authority says you ought to x, you have sufficient reason to x. Think, for example, of the legitimate authority of parents or teachers over children in their care. Merely *de facto* practical authority is the power to make claims which others take as sufficient reason for action. The difference between the two is that the claims in the first case are justified while in the second they are not. You may take the say-so of a *de facto* authority as sufficient reason for action, but it is not true that you have that reason.

Political legitimacy relates to the justification of the practical authority of political institutions and democratic legitimacy relates to the justification of democratic decision-making. So democratic legitimacy is an application of the concept of legitimate political authority to democracy. If political authority is legitimate, there is a right to rule. If democratic decision-making is legitimate, then the right to rule is jointly held by the members of the democratic collective. If democratic decision-making is legitimate, then the claims that can be associated with democratic decisions—the say-so of the democratic collective—give everyone sufficient reasons for action.

By legitimate epistemic authority I mean the right to make claims which give others sufficient reasons for belief.[3] If you legitimately hold epistemic authority

---

[2] I've discussed the two theories at length in Peter (2009).
[3] By belief, I here mean both full and partial belief; the reason for belief in question may thus either be a reason for a full belief or for adjusting your belief, e.g. for reducing your confidence in your original belief.

over p, then your claim that p gives me sufficient reason to believe that p. Just like in the practical case, we can also distinguish between legitimate and *de facto* epistemic authority. Epistemic authority is merely *de facto* if someone successfully pretends to have expertise that they in fact lack.

With these terms in place, we can now address the question: what is the relationship between democratic legitimacy and epistemic considerations? We can distinguish between two main approaches. According to the first, democratic legitimacy is independent of epistemic considerations and is established on grounds of the moral values embodied by democracy. That is the deep proceduralist scenario I mentioned earlier. Alternatively, epistemic considerations are at least one factor in the determination of legitimate practical authority. This is the approach epistemic democrats take. The main focus of my chapter is on the question how epistemic considerations should be brought to bear on the justification of practical political authority. In the next section, I will criticize the instrumentalist way of characterizing the relationship between epistemic considerations and the legitimate practical authority of democracy. In the rest of the chapter I will propose an alternative—proceduralist—way of characterizing this relationship.

2. Epistemic Instrumentalism

Variants of epistemic instrumentalism are popular among epistemic democrats today. According to them, the epistemic quality of the decisions made justifies the authority of democracy or is at least one justificatory factor. Here is a typical characterization of the epistemic conception of democracy:

> For epistemic democrats, the aim of democracy is to "track the truth." For them, democracy is more desirable than alternative forms of decision-making because, and insofar as, it does that. One democratic decision rule is more desirable than another according to that same standard, so far as epistemic democrats are concerned.
>
> (List and Goodin 2001, 277)

This characterization allows for different conceptions of democratic legitimacy, depending on how much weight is given to the outcome of decision-making relative to the decision-making procedure itself. One possible conception is what David Estlund has called the "correctness theory" of political legitimacy. On this conception, a political decision is legitimate if and only if it is the correct decision. With regard to the legitimacy of democracy, it says that democratic decision-making is conducive to political legitimacy to the extent that it successfully tracks a procedure-independent truth.

But making political legitimacy dependent on the correctness of the decisions presupposes a right to make claims about what the correct decision is, as without the possibility of judging what the correct decision is, it remains indeterminate whether a decision is or is not legitimate. So, on one way of making sense of the correctness theory of political legitimacy, someone, or a small group of people, must be holding the right to make claims about which democratic decision is correct and, as such, legitimate.[4] And this right to make claims about what ought to be believed derives from procedure-independent facts or objects or truths. Another way of saying the same thing is that the correctness theory of political legitimacy only works if there is third-personal epistemic authority about the matter to be decided—if someone, or a small group of experts, holds a right to make claims about what the correct decision is that derives from a procedure-independent truth. But this way of interpreting the correctness theory gives rise to a problem. For any area of decision-making where there is a procedure-independent right to make claims about what the correct outcome is, democratic decision-making is either redundant or it needs to be defended on other grounds.

The availability of third-personal epistemic authority presents the correctness theory of democratic legitimacy with the following *authority dilemma*: if practical authority is justified on epistemic grounds, then legitimate practical authority is non-democratic. If, on the other hand, the practical authority of democracy is to be legitimate, it must be justified on non-epistemic grounds. In other words, for those areas of decision-making where there is third-personal epistemic authority, we either follow those who know what the correct decision is, in which case our decision-making is not democratic, or we insist on democratic decision-making, in which case we can't defend the legitimacy of democracy on epistemic grounds but must defend it on purely practical grounds. What the authority dilemma shows is that this way of conceiving of the relationship between legitimate democratic authority and epistemic authority is self-undermining.

Can the epistemic instrumentalist defence succeed if we adopt Estlund's (2008) proceduralist alternative to the correctness theory of democratic legitimacy? The conception of legitimate democratic authority that he proposes retains the truth-tracking aim for democratic decision-making, while putting weight on democratic procedures as well as their outcomes. As he characterizes democratic legitimacy (2008, 98), it requires that the democratic decision-making procedure "can be held, in terms acceptable to all qualified points of view, to be epistemically the best (or close to it) among those that are better than random". The thought is that if the democratic decision-making procedure satisfies this criterion, its

---

[4] I shall later discuss an alternative way of cashing out the correctness theory.

outcomes, whether correct or not, are legitimate. In other words, correctness does not directly determine the legitimacy of decisions made; it only influences the legitimacy-generating potential of democratic decision-making procedures.

This is a step in the right direction, but I don't think this conception can avoid the authority dilemma, at least not if we follow Estlund's interpretation of his requirement of democratic legitimacy. Much hinges on what the qualification that the procedure should be held to be epistemically the best, in terms acceptable to all qualified points of view, is supposed to entail. I see two possibilities.[5] According to the first, what is acceptable to all qualified points of view (however they are understood) *constitutes* what is epistemically the best procedure. In other words, what is epistemically best is defined in terms of what is acceptable to all the participants. I don't think that Estlund has this interpretation in mind, as it would clash with the idea that democratic decision-making should track a procedure-independent truth. This leaves the second interpretation, which identifies the qualified points of view as those that acknowledge the independent—third-personal—epistemic merit of a decision-making procedure. On this interpretation, the procedure that is held to be epistemically the best is the one which best tracks a procedure-independent truth, as identified by those who either hold third-personal epistemic authority on the matter or who are prepared to defer to it.

This second interpretation of the relationship between the legitimate practical authority of democracy and epistemic considerations again presupposes third-personal epistemic authority and, as such, leads straight back to the authority dilemma. For Estlund's conception of democratic legitimacy to have any judgmental bite, there must be a procedure-independent right to make claims about which decisions are correct and which decision-making procedure is most likely to produce correct decisions. Interpreted in this way, the conception presupposes third-personal epistemic authority about the correctness of outcomes and about which decision-making procedure can best approximate it. But if there is such a right, democratic decision-making once again appears either redundant or, if it is not redundant, then its value must be non-epistemic.

If my argument so far is correct, it shows that the attempt to defend democracy from a third-personal epistemic standpoint is self-undermining. Third-personal epistemic authority can ground non-democratic forms of decision-making, but it cannot establish the legitimacy of democracy.

---

[5] Gaus (2011) makes similar points in his discussion of how to "test" whether democracy is the epistemically best procedure.

Which way forward for defending democracy? The authority dilemma suggests two possible strategies. The first is to drop the epistemic strategy altogether and defend democracy on purely practical grounds. As mentioned above, this is not the strategy I shall pursue here. The other is to find an alternative epistemic defense of democracy, one that is not affected by the authority dilemma. This is the strategy I shall pursue here.

But before I can do that, I need to address a potential objection to my argument so far. I have said nothing about the popular epistemic defense of democracy based on the Condorcet Jury Theorem (CJT) and related results. The objection is this: a defense of democracy that rests on the CJT and related results is not vulnerable to the authority dilemma, as these results show that we can rely on democracy to discover the correct decision.

The CJT says, roughly, that if there are two alternatives—where one is the correct choice and the other the wrong choice—and if every member of a democratic collective is more likely to make the correct choice than the wrong choice and they vote independently of each other, then the majority is also more likely to make the correct choice and the probability that it makes the correct choices increases rapidly with the size of the collective. If the relevant conditions obtain, the CJT shows that larger groups are more likely to make correct decisions than smaller groups or individuals; a democratic collective may even outperform a group of experts. The CJT can thus be used to defend democratic decision-making on the basis of the claim that—under certain conditions—epistemic authority is appropriately held by a large collective. Note that the CJT is only relevant for the process of aggregation, not for deliberation. It highlights features of large-scale aggregation of votes, where votes are understood as expressing beliefs about the correct decision. It does not cover deliberation, that is, the process of exchanging reasons for belief or action.

There has been much discussion of the likelihood that the conditions under which the CJT holds are ever met and, on that basis, whether it can be used as part of a defense of democracy.[6] But I shall not go into that discussion, as my purpose here is neither to criticize nor defend the CJT and its applicability to democracy. I'm interested in the question what an argument from the CJT would imply for the relationship between epistemic considerations and the legitimate practical authority of democracy.

The answer to this question depends on how the CJT is invoked to explain why an appropriately made democratic decision gives individuals sufficient reasons to uphold it. The CJT may be invoked as part of a correctness theory of political

---

[6] See List and Goodin (2001) and Dietrich and Spiekermann (2013) for recent discussions.

legitimacy. In this case, it is, as before, the correctness of the decision, not the democratic decision-making process, which generates legitimacy, and the authority dilemma looms. Or it may be invoked as part of an argument which shows that democratic decision-making is the most likely decision-making procedure to lead to correct decisions. In this case, too, the authority dilemma will reappear if the argument presupposes knowledge of what the correct decisions are.

But if it can be avoided in this second case, it is because the CJT is invoked in an argument that starts from circumstances in which we don't know what the correct decision is, that is, when the third-personal standpoint is unavailable. Understood in this sense, an argument from the CJT is compatible with the thesis that I want to defend here, namely that there is scope for democratic decision-making when there is no procedure-independent epistemic authority in the relevant area of decision-making. Unlike the arguments that end up in the authority dilemma, this kind of argument links the legitimate practical authority of democracy not to its instrumental role in reaching correct decisions, but to procedural epistemic considerations. It shows that, under certain conditions, the claims made by individuals or small groups of experts lack the epistemic authority that the democratic decision-making process can claim. In this case, our reason to defer to a democratically made decision is not because it is correct or likely to be correct, but because there is no procedure-independent way for assessing claims about what the correct decision is. If the conditions apply, then the CJT shows that epistemic authority is appropriately held jointly and exercised through a democratic decision-making process.

The argument I want to develop in this chapter parallels this kind of argument from the CJT. I shall focus on the deliberative component of democratic decision-making, rather than the aggregative component to show that the deliberative procedure itself may have epistemic value.

3. The Procedural Epistemic Value of Deliberation

The argument I develop in this section parallels the argument from the CJT just sketched in the sense that it also shows that there is a prima facie case for democracy when there is no procedure-independent way of assessing epistemic authority. I shall focus on the deliberative component of democratic decision-making, rather than the aggregative component. My aim is to show that the deliberative democratic procedure itself, that is, the process of exchanging reasons with others and of adjusting one's beliefs in response to the claims made by others, may have epistemic value—above and beyond the value of making correct decisions.

I understand procedural value in contrast to instrumental value. Deliberative decision-making has instrumental epistemic value if it leads to more accurate beliefs among the participants—for example, through comparing evidence and opinions and responding to the evidence and opinion that others present—and/or to correct or more accurate decisions. It has instrumental epistemic disvalue if it hinders the formation of accurate beliefs and/or the making of correct decisions. Epistemic instrumentalism reduces the epistemic value of democratic decision-making to its contribution to epistemic ends such as accuracy or truth.

By contrast, deliberative democratic decision-making has procedural epistemic value if its epistemic value does not reduce to the epistemic value of the outcome. I shall focus here on the procedural epistemic value of the deliberative process, not of decision-making as such. If public deliberation has procedural epistemic value, it has epistemic value even in cases where its effect turns out to be that it has diminished the accuracy of the beliefs of the participants. This may sound paradoxical, but I will show that it is possible and of significance for an epistemic defense of democracy.

Procedural value can take different forms. Rawls has identified the following three main forms: pure, perfect, and imperfect proceduralism.[7] With pure proceduralism, the procedure is necessary and sufficient for the value of the outcome. There is thus no procedure-independent standard for what counts as a good outcome. With both perfect and imperfect proceduralism, there is a procedure-independent standard. With perfect proceduralism, the procedure is necessary to realize a good outcome. With imperfect proceduralism, the procedure is necessary to approximate a good outcome, but it may fail to realize the outcome that the procedure-independent standard envisages. For example, in a trial there is a procedure-independent standard for a good outcome: convict the accused if guilty and don't if not guilty. The trial is necessary to approximate such good outcomes. But justice may be served even if an appropriately conducted trial ends up making a decision which, with hindsight or from some other vantage-point of privileged information, turns out to be wrong.

As I will explain below, the procedural epistemic value of deliberation takes the form of imperfect proceduralism. On that interpretation of procedural epistemic value, accuracy of belief remains the intrinsic epistemic value and sets a procedure-independent standard for evaluating deliberation. But what distinguishes imperfect proceduralism from instrumentalism about epistemic

---

[7] Rawls (1971, 85); he introduced them with regard to the value of justice, but the distinctions apply more generally.

value is that the former captures the situation in which the procedure has value because there is no procedure-independent access to the correct outcome. The deliberative procedure thus gains its value from being a proxy for good epistemic conduct in situations where it is not possible to appeal to the procedure-independent standard directly to assess this conduct. And this value of the deliberative procedure does not reduce to the epistemic value of its outcome as assessed by a procedure-independent standard.

The idea that deliberation has procedural value is familiar from the practical context. Rawls, for example, takes reciprocity to be a fundamental procedural value that shapes the content of the claims of justice we have on each other. In moral philosophy, Charles Larmore (2008) focuses on equal respect and Stephen Darwall (2006) on mutual accountability. But what I'm claiming here is that procedural values play a role in the purely epistemic dimension of deliberation as well.[8]

How should we make sense of the idea that deliberation has procedural epistemic value? It turns out that the epistemology of disagreement shows that, under certain circumstances, epistemic agents have reason to engage in deliberation with each other and adjust their beliefs in direction of each other. This value of deliberation does not reduce to the value of the outcomes of deliberation, that is, it is, to some extent at least, independent of its contribution to the accuracy of belief.

The relevant circumstances arise only with certain types of disagreement. Specifically, they arise with disagreements among epistemic peers that are persisting in the sense that deliberation fails to reveal evidence that one party left unexamined or a mistake one party made in the interpretation of the available evidence. An epistemic peer is someone who you take to be equally likely to make a mistake. This is a weak definition of what it means to be a peer, since it only takes the form of an all-things-considered criterion and doesn't involve any input conditions such as equal formal qualifications or equal computational abilities. On this definition of peers, deliberation among parties who consider each other peers can occur not just in the context of academic or expert inquiry, but in any small or large social collective, for example, on issues which are too wide-ranging and complex for anyone to count as an expert, or when relevant information is dispersed across all deliberative parties. It turns out that, in those circumstances, each party to the deliberation has reason to adjust their beliefs in direction of each other.

---

[8] The following passages borrow from Peter (2013).

But before I go into more detail on this, it is necessary to address the objection that the epistemology of peer disagreements is not relevant for democratic theory, since the participants in democratic deliberation neither do nor should consider each other as epistemic peers. There are massive epistemic inequalities among the members of a democratic collective, and this renders the epistemology of peer disagreement unsuitable.

In reply to this objection, let me first concede that these epistemic inequalities are often both massive and justified. You might very well know a lot more about certain subjects than other people, and may validly claim epistemic authority on the matter. I'm not suggesting that the members of a democratic collective generally are epistemic peers, nor that they generally ought to regard each other as such. The question I'm interested in is when deliberative democracy—some form of rule by all—is appropriate. And the claim that I intend to defend is that when there is a disagreement among parties that do, in a non-crazy way, take each others as peers, then deliberative democratic decision-making may be appropriate. By contrast, when some parties hold legitimate epistemic authority over an issue, the epistemic point of view supports a form of epistocracy—rule by experts.

Secondly, note that the notion of peerhood I'm using is compatible with all sorts of epistemic inequalities. You might consider someone a peer—with regard to a certain set of issues—even though your formal qualifications, your computational abilities, or your access to information differ. In addition, this notion of peerhood is domain-specific. It is possible to accept someone as an epistemic authority on some questions but consider this person a peer with regard to certain political matters at the same time.

Thirdly, note that while you judge your peers, by definition, to be equally likely to make a mistake in a particular situation, this doesn't mean that two peers always perform equally well. Sometimes you have information about the circumstances of the disagreement that makes it appropriate for you to discount their judgment, at least to a certain extent. For example, of two scientific colleagues who compare their different conclusions about the validity of a hypothesis, one may have double-checked the data and the calculations and asked an assistant to do the same, while the other was pressed for time and admits that he only ran what he was given through an off-the-peg computer program. In a context of social deliberation, imagine a case of committee work. Suppose that, although all participants regard each other to be equally able to take up the available evidence, some may have carefully thought about the implications of the evidence presented, as manifested by the quality of the arguments they can offer in support of their beliefs, while others respond with a gut reaction. If there is such information which suggests that a peer is not performing well, you are entitled to discount their view.[9]

So much for the negative defense of using the epistemology of peer disagreements as a starting-point for understanding democratic deliberation. It shows that my claim is not a very strong one.[10] But I haven't said anything positive yet about how the epistemology of peer disagreements might help. I now want to get to that.

Consider the following case. An expert committee prepares a policy together with relevant politicians, for example, a new minimum-wage policy. Suppose there is a disagreement on the question whether all the evidence, appropriately interpreted, supports the policy or not. Will the policy significantly increase unemployment? This would speak against the policy. Or will it not, and have other beneficial aspects? This would speak for the policy. Suppose the disagreement is not just between experts and politicians; the experts disagree among themselves. Also suppose the disagreement is not the result of one party not being able or willing to consider the available evidence, draw appropriate conclusions from it, and so forth. What we have here is a persisting disagreement among parties that do take each other as peers. Now suppose that the disagreement isn't limited to this particular committee, but is mirrored in similar committees, in parliament, in academia, in the media, and elsewhere. And it is also mirrored in debates among friends and family members. Suppose you have looked at the available data and the available arguments by the experts and formed the belief that the proposed minimum-wage policy will significantly increase unemployment. I have also looked at the data and the arguments and formed the belief that the policy will not significantly increase unemployment (and will also have very

---

[9] You might even be entitled to discount their view completely, as the following case illustrates (adapted from Christensen 2007). Suppose you and I regularly have dinner together at a restaurant and we always split the bill. Neither of us has any problems with mental arithmetic, so the decision about how much we each owe is usually straightforward. One evening, however, I claim we each owe £26 and you claim that we each owe £28. In this case we each have reason to suspend our belief about how much we each owe. But now suppose that instead of claiming that we each owe £28, you claim we each owe £280, way more than the total bill. Even though I consider you a peer in this regard, something has clearly gone wrong and I'm entitled to stick to my guns without giving any weight to your belief.

[10] In fact, it may be weakened further. The epistemological literature currently focuses on peer disagreement, and my argument is based on initial findings of this research and is thus somewhat hostage to that literature. As will become clear later, what is key to my argument is the possibility that you find yourself in a disagreement with someone to whom you are attributing some epistemic credibility and to whom you owe some sort of response. The peerhood assumption imposes symmetric credibility. But that assumption is not necessary, it seems to me. As long as all of the parties attribute some epistemic credibility to each other, then, everything else equal, some sort of response may still be required of each, and that would be sufficient for my argument to go through.

positive effects on working conditions). Through deliberation we become aware of our disagreement, and neither can demonstrate that the other obviously made a mistake.

I want to claim that in a case like the minimum-wage case, what we have is a peer disagreement writ large, extending, potentially, to the entire democratic collective. What is characteristic of this situation, from an epistemic point of view, is that each side of the disagreement deserves some weight for the claim it makes, and there is no vantage-point from which the disagreement could be resolved.

What is the appropriate response to such a disagreement? I follow the majority view here, which is that it is appropriate that both sides to a peer disagreement diminish confidence in their original beliefs.[11] Different epistemological theories give different accounts of why this is so. According to the Total Evidence View (Kelly 2010), if a disagreement with a peer receives any weight, it is as a piece of evidence that a rational epistemic agent needs to consider together with the other available evidence. On the rivaling Conciliatory Views (Elga 2010; Christensen 2011), the disagreement with a peer gives you a reason to adjust your belief that is, at least to some extent, independent of the reasons that you have had to form your original belief. On this view, you have done the best you could, given the evidence, and so on. But that was apparently not good enough, as your peer has formed a different belief. You now need to put the reasons you have had to form your original belief to the side and respond to the situation that the disagreement has revealed.

There are significant differences between these two main views, but for our purposes here these differences do not matter, as both theories concur that, absent independent information about their respective ability or willingness to perform in the particular instance, peers need to adjust their beliefs in direction of the other party. Both accept that the fact that someone you *ex ante* regarded as a peer now disagrees with you is not sufficient for you to dismiss their opinion off-hand. Some sort of response is required.

This shows that there are circumstances in which there are good epistemic reasons for epistemic peers to be responsive towards each other's claims and to consider some revision of their original beliefs on the basis of these claims. I have called this the accountability thesis about the epistemic value of deliberation

---

[11] Some (e.g. Kelly 2005) have argued that peer disagreements do not require a response at all, by claiming that what justifies someone's belief is their response to the available evidence and by denying that information about the beliefs of peers carries independent epistemic weight. But many have rejected this "steadfast view", and I follow them here.

(Peter 2013). Epistemic peers, in these circumstances, are mutually accountable to each other, in addition to being accountable to the truth they seek. In the relevant circumstances, your claim (that the minimum-wage policy will increase unemployment) gives me a reason to revise my belief (that it will not) and vice versa. What the accountability thesis captures is that there are situations in which we are doubly accountable. There is, on the one hand, the familiar accountability to truth. But there is, on the other, also an often overlooked accountability to epistemic peers. When we find ourselves in a persisting disagreement, where neither party can demonstrate that the other is simply wrong, what gives each of us reason to adjust our believes in direction of the other are the claims we both made, not just first-order evidence about the fact or truth in question.

What is more, in those circumstances neither of us can validly claim third-personal epistemic authority. Insofar as there is any appeal to epistemic authority, it takes on a second-personal form. To see the point, contrast the deliberative situation here with a case of testimony. In testimony, we are also considering an epistemic relationship—the relationship between the testifier and the addressee. This relationship is hierarchical, however, not democratic. If the relevant conditions for successful testimony are met, your testimony gives me a sufficient reason for belief. My reason for belief is thus second-personal; it derives from your claim. But your authority to make claims that give me reasons for belief is not; it derives from your accountability to the truth and is third-personal. In the case of peer disagreement I'm considering here, each of us has a second-personal reason to adjust our beliefs. We're both accountable not just to the truth that we both seek, but to each other as well. Good epistemic agents enter this relationship of mutual accountability and let it be a—second-personal—source of reasons for belief that is, at least to some extent, independent of truth as a source of reasons for belief. To this extent, epistemic authority, too, takes on a second-personal form. My right to make claims that give you reason for belief depends on your accountability to me as someone you regard as a peer, and vice versa.

I'm not denying that accuracy or truths are the sole epistemic ends. But what my argument shows, I hope, is that it is possible to value the deliberative democratic procedure in non-instrumental fashion. This opens the door to an alternative to the epistemic instrumentalist defense of democracy. When peers cannot agree on what belief is warranted, and nothing suggests malperformance, then at least some reasons for belief stem from a relationship of mutual accountability between them. But mutual accountability is a procedural consideration. Its epistemic value cannot be captured by a purely instrumental approach because, if the relevant conditions apply, my reason to adjust my belief in your direction obtains independently of whether the adjustment makes my belief more accurate.

4. The Legitimate Authority of Democracy

The alternative to the epistemic instrumentalist defense of democracy that I'm proposing is this. There is an epistemic case for democracy when the epistemic circumstances are such that there is a peer disagreement writ large on a particular issue, one that cannot be resolved by appeal to third-personal epistemic authority and one that places the members of a democratic collective in a relationship of mutual epistemic accountability. In this final section, I want make a few brief comments on how this claim is to be understood.

The situation I have focused on is one in which deliberation has established that no participant can legitimately make decisive claims about what others ought to believe about what the correct decision is. The necessary adjustment of belief may have led them all to suspend belief. Or it may have led to a reasonable disagreement, that is, a situation in which parties to a disagreement hold incompatible beliefs, but each has some justification for holding the belief they do.

If a decision is needed, then some form of aggregation may be required in these circumstances to reach a decision. While a lot more would have to be said about when and how to aggregate, something I can't do here, what we are beginning to see is how democratic decision-making, understood as a combination of deliberation and aggregation, has its natural locus in a situation in which the reach of epistemic authority is insufficient to determine what the correct decision is.

Note that my aim in this chapter is a modest one. I'm not trying to establish the set of requirements which is jointly sufficient for democratic legitimacy. Instead, I'm interested in how to identify the epistemic circumstances in which deliberative democratic decision-making is potentially legitimate. In other words, I'm concerned with circumscribing the logical space in which democracy belongs.

The view that I have outlined suggests that deliberative democratic decision-making has its appropriate space in situations where disagreements cannot be resolved by appeal to third-personal epistemic authority. When they can, then there is no epistemic basis for democracy. To the contrary. If legitimate practical authority is premised on epistemic considerations, and if there is third-personal epistemic authority, the decision should be made by those who know what the correct decision is. If democracy has any value at all in those epistemic circumstances, it is for non-epistemic, practical reasons. But if these disagreements cannot be resolved, then the decision has to be made on other grounds rather than by appeal to third-personal epistemic authority. Epistemic considerations, in this case, help with identifying the scope for democratic decision-making and impose certain conditions on the deliberative democratic decision-making process.

What is the connection between the procedural epistemic value of deliberation and democratic legitimacy? The view I've outlined supports a combination of imperfect proceduralism about epistemic value with pure proceduralism about the legitimate practical authority of democracy. The epistemology of disagreement shows that, in certain circumstances, the epistemic value of deliberation doesn't reduce to whether it produces more accurate beliefs, but depends on whether or not the process itself is conducted appropriately, that is, on respect of the relationship of mutual accountability between the participants. Since an appropriately conducted process of deliberation is a proxy for aiming at forming accurate beliefs in circumstances where disagreement reveals a difficulty with determining directly what the correct belief is, epistemic value in this case conforms to the structure of imperfect proceduralism.

In circumstances where there is no recourse to a procedure-independent standard for correct decisions, the legitimate practical authority of democratic decision-making can't depend on this standard. This suggests pure proceduralism about the legitimate practical authority of democracy. On this conception of democratic legitimacy, a decision is legitimate if and only if it has been made through appropriate deliberative decision-making procedures.[12]

Can the view I've outlined here avoid the authority dilemma? I think it can. The authority dilemma arises for epistemic instrumentalism about democratic legitimacy because of the tension between the legitimate practical authority of democracy and procedure-independent, third-personal, epistemic authority. The view I propose distinguishes between the imperfect proceduralism that shapes legitimate relations of epistemic authority and the pure proceduralism of democratic legitimacy. Epistemic considerations, on this view, influence the deliberative decision-making procedure, but they do not set a procedure-independent standard for democratic decision-making. Unlike forms of epistemic instrumentalism, it can thus explain under what epistemic circumstances it is appropriate to put practical authority in the hands of a democratic collective.[13]

---

[12] As such, the view I endorse here contrasts with Estlund's imperfect proceduralism about democratic legitimacy, as his view, but not mine, makes democratic legitimacy dependent on the epistemic quality of the outcomes of the democratic decision-making process.

[13] I have benefited greatly from comments and suggestions from participants at conferences, workshops, and research seminars in Basel, Belgrade, Geneva, Hull, Oxford, Rijeka, Stirling, and Zurich, and I'm very grateful to the organizers of these events. I particularly want to thank Yann Allard-Tremblay, Elvio Baccarini, Chris Bertram, Robin Celikates, Bruce Chapman, Tom Christiano, Rowan Cruft, Stefan Gosepath, Antony Hatzistavrou, Chandran Kukathas, Katrin Meier, David Miller, Snježana Prijić Samaržija, Henry Richardson, Ben Saunders, Kit Wellman, Jo Wolff, and Zofia Stemplowska. In addition, I want to thank Michael Brady and Miranda Fricker for detailed and very helpful comments on earlier drafts.

8

The Transfer of Duties
From Individuals to States and Back Again

Stephanie Collins and *Holly Lawford-Smith*

1. Introduction

We often make claims about the duties of states. One might think, for example, that Australia has a duty to legalize homosexual marriage, that Britain has a duty to spend less public money on the royal family, or that the United States has a duty to make healthcare affordable. Can such duties be explained in terms of individuals' duties? We think many—though not all—can. To make a case for this, we build a general and abstract analysis of how some collectives' duties are explained by individuals' duties, and apply it to states in a way that retains common-sense intuitions about states' duties. We then use that analysis to suggest a new sense of membership in a state. The analysis depends crucially on epistemic elements, including the beliefs and transfers of information required for reciprocal recognition and intentional participation; certain other bidirectional transfers of knowledge; and the required degree of justification for certain key beliefs.

There is surprisingly little literature on the possibility of a derivative relationship between particular duties of states and individuals. Much has been written assuming that states are moral agents and so can bear duties (e.g. Shue 1999; Reidy 2004; Barry 2005). Others have argued that states are moral agents, but

---

We are grateful to audiences at Collective Intentionality VIII (Manchester, 28–31 August 2012), the University of Essex, Nuffield College at the University of Oxford, the University of Sheffield, the University of Nottingham, and the University of Manchester; and for comments on earlier drafts, to Nic Southwood, Jonathan Quong, Federico Zuolo, Bob Goodin, Brett Calcott, and Miranda Fricker.

haven't commented on how those agents acquire or discharge duties, or who counts as a member (Goodin 1995, ch. 2; Erksine 2001; Wendt 2004). There is a growing literature on assigning duties of financial compensation to citizens, as a means of rectifying democratic states' wrongful acts (Thompson 2002; May 2006; Stilz 2011; Pasternak 2013). But rectifying wrongs makes up only a tiny subset of states' duties. What about their duties simply to pursue some good state of affairs, such as human rights fulfilment, climate change mitigation, or nuclear disarmament? And what about the question not of who should bear the costs of discharging these duties, given that there are such duties, but of how states acquire such duties in the first place? Our guiding questions in this chapter are: can we give an account on which a good number of states' duties are justified by and discharged through individuals' duties, and can this tell us anything interesting about membership in states? We give an affirmative answer to both.

These questions could easily be misread as belonging to the large literature on political obligation. There are well-known social contract accounts of the original transfer of individuals' rights in creating the state (Hobbes 1651 [1994]; Locke 1689 [2004]), just as there are various accounts of whether, and how, the state has authority over the individual (for an overview, see Dagger 2007). Our account may have implications for thinking about political obligation insofar as we suggest a criterion for membership, and one major theory of political obligation builds obligations out of membership (see e.g. Gilbert 1993). But we are not interested here in individuals' duties *in general* to obey the laws of a given state. Rather, we are interested in how any given duty of a state is derivable from particular duties of particular individuals, and how it will come to create moral duties (not necessarily legal ones) for those same individuals. It is a separate question whether the state's general authority over those individuals (or other individuals) is legitimate.

We begin in section 2 by laying out what we call 'the simple story'. The simple story starts with the fact that individuals sometimes discharge their duties by transferring them to collectives. The individuals' discharged duties generate new duties for the collective, and the collective discharges those duties by acting through its constituents, which involves distributing roles to them. (On individuals' duties in group contexts, see e.g. Feinberg 1968; Held 1970; Gilbert 2001; Lawford-Smith 2012; Collins 2013.) Constituents then incur yet more duties, related to their roles.

In section 3, we argue that this analysis can account for a good number of states' duties. We begin by mapping the general form of states as collective agents. We then explore the various components of the simple story as applied to states, and demonstrate its intuitive force as a model of many of states' duties. We close section 3 by addressing some complications that arise in applying the simple story to states. The result is that many, but not all, of states' duties can be accounted for by the simple story.

In section 4, we explain the various ways in which individuals and states can fail to properly discharge their duties in the way the simple story demands. We then find that the simple story produces some counterintuitive results regarding the constitution of a state's agency. In remedying these, we develop a novel and important sense in which individuals are members of states. This argument stakes out new territory within the growing literature on whose moral agency is implicated when the state acts (Parrish 2009; Stilz 2011; Pasternak 2013). We thus draw a bridge between theories of group duties, and theories of state membership, by using our theory of group duties to answer neglected questions about the normative basis of states' duties.

2. The Simple Story

To avoid rehearsing arguments given elsewhere (Pettit and Schweikard 2006; Pettit 2007, 2010; List and Pettit 2011; Collins 2013), we will simply assume that the only groups that can bear duties are those that have a shared decision-making procedure. (For contrary views, see Feinburg 1968; Held 1970; Wringe 2010.) Let us call these groups collective agents, or simply collectives. Each collective (and no non-collective group) has a procedure that takes inputs from some individuals, such as their beliefs, desires, and so on. It processes the inputs, resulting in its own beliefs and desires, and ultimately in a group decision and a distribution of roles to its constituents that are together sufficient for carrying out that decision. Collectives (and no other groups) can bear duties, since they (and no other groups) have their own group-level decision-making procedure.

We call our general account of collectives' duties 'the simple story'. It is presented visually in Figure 1.

The simple story has three 'nodes': {A}, {B}, and {C}. Each node is represented by a box. The second row in each box indicates the type of agent that has duties at

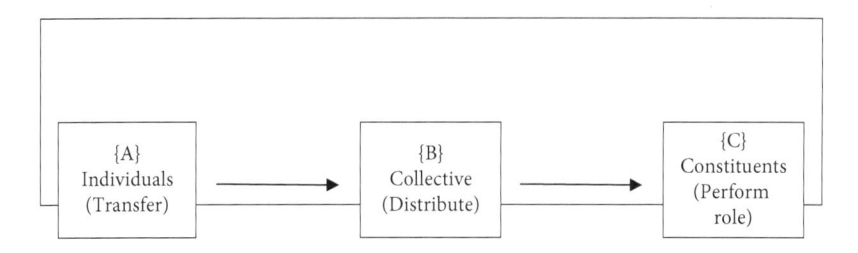

```
┌─────────────────────────────────────────────────────────────┐
│   ┌──────────────┐      ┌──────────────┐      ┌──────────────┐
│   │     {A}      │      │     {B}      │      │     {C}      │
│   │ Individuals  │─────▶│  Collective  │─────▶│ Constituents │
│   │  (Transfer)  │      │ (Distribute) │      │  (Perform    │
│   │              │      │              │      │    role)     │
│   └──────────────┘      └──────────────┘      └──────────────┘
└─────────────────────────────────────────────────────────────┘
```

Fig. 1. 'The simple story'

that node. In the bottom row is the method the relevant agents can use to discharge their duties at the relevant node (we say 'can' rather than 'must' because, as we will explain, there are sometimes multiple ways to discharge {A}- and {C}-node duties). At {A} the agents are individuals, and they can discharge (some of) their duties by transferring those duties across to {B};[1] at {B} the agents are collectives, and they in turn discharge the duties they have in virtue of the transfer from {A} by distributing roles to {C}; and at {C} the agents are constituents of the collectives at {B}, and they discharge their duties (usually, but not always) via the roles they have been assigned at {B}.

To illustrate, imagine an isolated, relatively poor community of a hundred people. The community is struck by an earthquake. Suppose that there is nothing any of them could have done—either individually or collectively—to foresee the earthquake or pre-emptively adapt to its effects. Most community members suffer no significant effects. However, ten people's worldly goods are completely destroyed. They are now homeless and penniless. According to most moral theories, the other community members each have a duty to do what they can, although not at disproportionate cost, to help the earthquake's victims.[2] By doing this (assume) they will maximize aggregate or average well-being, abide by categorical imperatives, do what the good person would do, make the worst-off better off, respect the victims' rights, and so on.

Suppose that, given their own poverty, each unaffected individual lacks the ability to do anything to significantly help the victims on his or her own. But imagine that each can, at not-disproportionate cost, take individual steps towards seeing to it that there is a collective that is able, at not-disproportionate cost and on her behalf, to provide the victims with subsistence goods until they are back on their feet. And assume that if each individual takes steps to ensure that such a collective exists, then such a collective will exist and will have a duty to provide the victims with the subsistence goods. Each individual can then discharge her duty to do what she can for the victims by signaling to the others that she is conditionally willing to take these steps ('I'll help bring about the capable collective if I reasonably believe you will'), and by taking the steps if the antecedent ('I reasonably believe you will') is true. These steps might include getting the others to help her create a collective, joining an established collective, working to change the capacities or functions of a collective to which she already

---

[1] As we will go on to explain, it's not strictly speaking the *duties* that get transferred between {A} and {B}, at least not in the particular form they took at {A}.

[2] For a useful discussion of moral demandingness and acceptable costs, see Goodin (2009), 6.

belongs, or simply checking to see that a collective to which she already belongs is pursuing that good on her behalf. Call these individual steps 'collectivizing'.

If the only way the unaffected individuals can help the victims is to collectivize, then this will be the only way to discharge their {A}-node duties. If the victims can be helped most efficiently if the individuals collectivize, then the individuals might be morally demanded to discharge their {A}-node duties in this way. However, if each individual is able to efficiently help the victims (at not-disproportionate cost) without a collective, then collectivizing is just one of many ways in which the individual might discharge her {A}-node duty. (For a roughly similar account of duties to form collectives, see Collins 2013.)

Discharging one's {A}-node duty by collectivizing requires more than just taking steps to ensure a capable collective exists. It also requires transferring one's {A}-node duty to the collective. Let us explain. Suppose the justification of Tui's {A}-node duty was that some good ought to be pursued—the good of 'the earthquake victims being helped.'[3] That general justification gives rise to specific duties for agents—whether individuals or collectives—to pursue that good. These duties are 'specific' in that their precise content depends on the agent's capacities. For Tui's specific duty to be passed to a collective, and thus for her to have discharged her {A}-node duty via the simple story, she must meet a particular epistemic condition: she must reasonably believe that a collective will pursue, on her behalf, the good of 'the earthquake victims being helped'—that its specific duty will be discharged in her name. Tui's discharging her {A}-node duty by collectivizing has four steps. First, she must do what she can (at not-disproportionate cost) to bring about a collective that can pursue the good of the victims being helped. Second, she must will that the collective pursue that good on her behalf. Third, she must communicate to the collective that she wills this (as we'll see, her communicating this to a state will require that she engage in political participation of some kind). Fourth, she must do what she can (at not-disproportionate cost) to bring about her own reasonable belief that the collective will, in fact, pursue that good on her behalf.

At first glance, it looks like what is transferred between {A} and {B} is Tui's specific duty to do what she can to pursue the good of 'the earthquake victims being helped'. She has literally 'passed' her duty on to someone (something) else. But for familiar reasons to do with 'ought implies can', Tui cannot have duties to do what she cannot do. So even though the good 'the earthquake victims being

---

[3] We hope this is fairly neutral between ethical theories. One might instead have 'Tui (and those who, on Tui's behalf, discharge duties with this justification) responding appropriately to the victim's claims', or similar.

helped' generates duties for both a single individual and a large collective, the specific duties that it generates will be different in content depending on the agent they accrue to. Tui does not pass on to the collective her specific duties, which are constrained by her individual capacities, but rather generates a new specific duty to pursue that same good, which is held by the collective and is constrained by *its* capacities. The collective's duty will demand different actions than Tui's duty, because the collective has different—much greater—capacities. This way of understanding what is transferred avoids the odd effect of a collective coming to have multitudes of specific duties to do things which are not sensitive to its own capacities.

In some cases, the collective would have incurred a capacity-relative duty to pursue the relevant good even if individuals had not transferred their general duties. But the transfer is important: the individuals' transferring of their duties constitutes an additional justification—over and above the value of the good to be pursued—for the collective's duty. In cases where the collective would have had a duty to pursue the good even without the individuals' transfer, the transfer overdetermines the collective's duty to pursue that good. The transfer is a sufficient—though not always necessary—justification for the collective's duty to pursue the relevant good. So it provides an explanation of the collective's duty. Additionally (and importantly for our later arguments), the transfer makes it true that when the collective discharges (or fails to discharge) its duty, it does so on the individuals' behalf.

Notice, though, that the collective might take itself to be pursuing a good on an individual's behalf, where she objectively has a duty to pursue that good at the {A}-node, but without it actually doing so on her behalf. In order for a collective to pursue a good on an individual's behalf, she must *will* that it does so and must *communicate* this will to the collective. If she does not will this, or if she does not communicate it, the collective isn't really pursuing it on her behalf. In such cases, she has not engaged in the duty transfer process with the collective. The collective's act of pursuing the good, and purporting to do so on her behalf, might serve to dissolve her {A}-node duty, but she has not thereby discharged her duty: if we imagine she has a moral ledger, she would not get a tick in the {A}-node duty row.

Note also that an individual might discharge his {A}-node duty by willing, communicating, and reasonably believing in the right way, even if the collective fails to so act. This is consistent with saying that the individual acquires a new duty when he discovers the failure. But his original duty was to take certain steps—namely, to will, communicate, and reasonably believe—which he did take. The original duty is thereby discharged, through the individual's meeting these epistemic conditions. (This will become important for the analysis in section 3, when we consider how individuals discharge duties by passing them on to states.)

At the {B}-node, there is a collective agent. The individual's transfer at the {A}-node has discharged her duty. It has also generated a new duty, held by the collective, with the same justification as the {A}-node duty: in the example, the justification was the good of the earthquake victims being helped. Thus the collective comes to have a duty to do what it can, at not-disproportionate cost, to pursue that good. In the example, the individuals' {A}-node duties amounted to signalling conditional willingness and acting if they reasonably believed the condition was met, and we supposed that the collective would be able, at not-disproportionate cost, to see to it that the victims have subsistence goods until they are back on their feet. So in this case the collective's duty will be to see to it that the victims have subsistence goods until they are back on their feet.

How does the collective discharge its duty? Collective agents act by having their constituents act, in just the way that individual agents act by having parts of themselves act. I start to dance a 1920s Charleston by lifting my right foot up while turning both ankles out, and then setting my right foot down while turning both ankles in. Collectives ensure subsistence goods for victims by distributing roles to constituents, such that if each constituent used their role in the right way, then the victims would have subsistence goods.[4] The collective 'distributes roles' when at least one individual uses the collective's decision-making procedure to distribute roles.[5] For example, following the conditions for group agency in Pettit and Schweikard (2006), suppose the ninety individuals unaffected by the earthquake set up a governing board of five; express their intentions to have the board's decisions bear on their own; establish their decision-making procedure as 'consensus of the board'; establish the goal of their decision-making procedure as 'giving the earthquake victims subsistence goods'; and those nominated as board members agree on the roles jointly sufficient for achieving that goal, and communicate those roles to the constituents who must use them in pursuit of the goal. When the board has distributed the relevant roles, the collective has done its duty.

Finally, at the {C}-node, we have individual constituents of the collective and their duties to use their roles in such a way that the good is pursued. In the earthquake example, a role might be 'housing two victims' or 'baking ten loaves

---

[4] 'Constituents' are individuals who meet the relevant conditions for collective agency. We say 'constituents' rather than 'members' because later we will develop a conception of membership that is stronger than the concept of 'constituent' that we are using here.

[5] 'Roles' isn't meant in any deep sense: it may be equivalent to 'actions' (the role may be to perform an action). Furthermore, we talk about the distribution of roles because we're focused in this chapter on collective actions rather than collective decisions, but if the collective's goal is only to produce a decision, then the roles will have mental actions rather than physical actions as their content.

of bread'. Different constituents can be given different roles based on their various capacities, and their all-things-considered role-related moral duty holds only if the cost of performing it is not disproportionate.

At the {C}-node, instead of viewing the group as one entity discharging duties as a unitary agent as we did at {B}, we now consider the constituents' duties individually. This means that when we consider a given constituent's duties, we cannot simply assume that others will perform their roles (as the collective did when it was assigning the roles). A constituent's duty is not an isolated duty simply to perform her role. Instead, because the {C}-node duties derive from the {B}-node duties, the individuals at {C} have duties to use their roles in such a way that they do what they can (at not disproportionate cost) to see to it that their collective does its duty, that is, pursues the good in question (insofar as it can at not-disproportionate cost). An individual's using her role for this purpose might require more than just performing the role the group gives her. For example, Kea's using his role to see to it that the collective helps the victims might require not just (say) baking bread, but also motivating other constituents to perform their roles, and maybe even performing their roles for them if they fail (Lawford-Smith 2012). So the constituents may be required to help others do their bits. This is all part of their duty, which is 'use your role in the collective to see to it that the collective does its duty'. Constituents are jointly and severally responsible for the collective's doing its duty, because the duty is something they are discharging together, as a collective.

Although the constituents discharge the collective's duty together, this should not be taken to automatically imply anything about their psychology. Individuals sometimes intend, plan, wish, and so on, not just on their own behalf, but on the behalf of a group they partly constitute, whether or not that group is a collective agent (for accounts of this, see Bratman 1999; Pettit and Schweikard 2006; Tuomela 2006). Our simple story is neutral on the question of whether—at any of the three nodes—individuals 'put themselves in the group's shoes', so to speak. It is possible for the simple story to run with or without such 'we-thoughts'. If it does not include them, then each individual thinks only about what he or she has a duty to do, at each node. At {A}, this is willing and communicating; at {B}, this is ensuring that at least one constituent uses the collective's procedure to distribute roles; at {C}, this is using her role with a certain end in mind. The collective can discharge its duty without anyone conceiving of themselves as 'in its shoes'.

Nonetheless, it might be useful or common for an individual engaged in the simple story to think in we-terms—to think not just about doing her own duty, but about the collective doing its duty. These kinds of thoughts might be particularly useful at {B}, where at least one individual must distribute roles with a view to a certain outcome. We-thoughts might also be helpful at other nodes. For example, at {C} they might help an individual to better understand the purpose of her role-related duty. But nothing in our account of duty transfer hangs on this psychological question.

The psychological question to one side, there remains a metaphysical question about the relationship between the individuals and the collective. Is the collective somehow 'more than the sum of its parts' or 'something over and above its constituents'? We think not. An assumption of our model is that, ontologically speaking, a collective is nothing more than a set of individuals arranged in a certain way, for example, meeting something like the conditions for collective agency presented in Pettit and Schweikard (2006), 33. If this assumption were false, the simple picture would look quite different: for one thing, the duties at {B} would not be analysable in terms of individuals' duties. Those who endorse metaphysically non-reductionist views of collectives should view our project as an attempt to make sense, in metaphysically reductionist terms, of certain duties of collectives in general, and then of states in particular. If that project succeeds, we will not have shown metaphysical non-reductionism to be false, but we will have vindicated metaphysical reductionism's ability to explain the justification of collective's duties.

3. Application to States

State agents

We will now apply the simple story to states. This application is not entirely straightforward, because the size and complexity of states introduces epistemic problems that we assumed away in the earthquake example.

We understand 'states' in much the way described by the Montevideo Convention on the Rights and Duties of States (1933). The Convention describes states as having: '(a) a permanent population; (b) a defined territory; (c) government; and (d) the capacity to enter relations with other states.'[6] Some might be tempted to interpret claims like 'Australia has a duty to legalize homosexual marriage' as 'the Australian government has a duty to legalize homosexual marriage'—here we're not interested in the narrow understanding of states as governments, but in a very broad understanding of states as collective agents that include anyone whose agency is implicated in the states' actions.

---

[6] According to Jasentuliyana (1995), 20 and Shaw (2003), 178, these criteria remain generally accepted.

The claim that states have group-level decision-making procedures—and are therefore collectives—is plausible, even if controversial.[7] States meet the conditions for agency given in, for example, Pettit and Schweikard (2006). States have a range of complex decision procedures, which systematically produce a range of beliefs, desires, and goals, and a distribution of individual roles for achieving those goals. Many people (implicitly or tacitly) intend to have the goals, procedure, and role distributions bear upon their own decision-making goals and procedures. A state distributes roles to constituents on the basis of its goals, which are a complex function of constituents' goals. (Recall that 'constituent' here means those individuals who meet whatever are the correct conditions for being a member of a collective agent. Under Pettit and Schweikard's model, the key condition is that the individual intends to have the collective's decisions bear on their own decisions.) To be sure, states' decision procedures often turn constituents into adversaries and some constituents arbitrarily have more sway over the procedure than others. States' decisions are not reached by anything remotely resembling consensus. Moreover, not all of the procedures are explicated and they often change in *ad hoc* ways. Yet the result of these processes within the collective is a set of goals, a set of individual roles for achieving those goals, and a distribution of the roles among individuals: results produced not by one constituent—or by the conjunction of each constituent's independent processing—but by the constituents as a system. The will of the state is a complex function of the will of its constituents. The actions of those with roles under the state's decision procedure partly constitute actions of the state when the role-holders use their role to help discharge the state's duty.[8]

The nodes in the state story

The simple story can be used to shed light on a large number of states' duties. Consider a state's duty to mitigate climate change, promote fair trade, or respect and protect human rights on a global scale. What reasons might we give for why states have these duties? We might point to the state's relationship to the individuals it represents: that the state owes it to those individuals to pursue those goods. But often, the individuals that states represent will not be the

---

[7] Erskine (2001) and Wendt (2004) agree, each using slightly different criteria for collective agency.

[8] On this, our account agrees with international law. The *Responsibility of States for Internationally Wrongful Acts* (<http://untreaty.un.org/ilc/texts/instruments/english/draft%20articles/9_6_2001.pdf>, ch. 2, article 5) states: 'The conduct of a person or entity . . . which is empowered by the law of that State to exercise elements of the governmental authority shall be considered an act of the State under international law, provided the person or entity is acting in that capacity in the particular instance.'

intended beneficiaries of states' duties: for climate change, it will be future generations; for fair trade and global human rights, it will be people all over the world (including, but not limited to, those a given state represents). It then seems odd to say that the state owes such duties to the people it represents—unless what states owe to the people they represent is to pursue some goods on those individuals' behalf.

Moreover, it is independently plausible that individuals have duties at the {A}-node that relate to pursuing the goods of climate change mitigation, fair trade, and so on. If we could effectively pursue these goods on our own, then, according to a wide range of ethical theories, we would have a duty to do so. But we cannot effectively pursue them on our own. You, as an individual, cannot have duties to bring about these goods, because you, individually, are unable to do so. But you can have a duty to do what you can, at not disproportionate cost, to pursue them. There are multiple ways to discharge your individual duties to pursue these goods, and often you will be required or will choose to do other than just transfer them to the state. In pursuing the good of avoiding dangerous anthropogenic climate change, for example, you might switch to a vegan diet or give up air travel. But you might also choose to transfer your duty to your state: to bring about your reasonable belief that the state has, first, taken on your duties and, second, is giving relevant roles back to you and other constituents. You can discharge your duty to pursue that good by doing what you can to make your state such that it can pursue that good; willing that the state pursues it on your behalf; communicating this to the state; and forming the reasonable belief that the state will, in fact, pursue it on your behalf. This is all just as the community members unaffected by the earthquake did in the initial example.

What specific actions do these four steps involve? Take the climate change mitigation duty at the {A}-node. Most states are able to pursue the good of climate change mitigation to some extent. (Note that an ability to do something does not entail a willingness to do it.) If one's state does not pursue it, then this first step might require anything from awareness-raising, to rallying for a new government department, to (in extreme cases) staging a coup.[9] The second step—willing that the state pursue climate change mitigation on your behalf—requires a simple mental action. The third step, communicating your will to the state, will require political participation. In democracies with the appropriate checks and balances, it will require voting in a certain way. In democracies and elsewhere, it might require protesting, signing petitions, publicly declaring one's views, or so

---

[9] Of course, whether one is all-things-considered obliged to do this depends on one's capacities, proportionate cost considerations, and the possibility that one's actions won't be futile.

on. Finally, bringing about your reasonable belief that the state has taken on your duty might just require regularly checking that the state already is pursuing that good on your behalf (for example, by following the political news). Or it might require even less than this: if it is epistemically reasonable for you to simply trust that your state is doing as it ought, then you might be able to secure a reasonable belief simply by trusting. If you reasonably believe that the state has not taken on your duty, then bringing about your reasonable belief that it has might require rallying your government representative for more aggressive climate policy, joining a green movement, raising others' awareness, and so on. In this way, the {A}-to-{B} transfer requires that individuals get themselves into a particular epistemic position.

Our duties at the {A}-node are constantly being discharged as we keep willing the transfer, communicating this to our states, ensuring that our beliefs that our states have taken up our duties are reasonable beliefs, and working to reform the state if we see that such a belief is not reasonable, in order to make it so. In extreme cases, discharging an {A}-node duty might require attempting to join a state (for example, by emigrating there) or attempting to form a new state (for example, by joining a secession movement).

As we said, there are epistemic problems for the simple story's application to states. One might wonder about, for example, a climate change sceptic, who believes there is no {A}-node duty to pursue the good of climate change mitigation. Assume his belief that there is no such duty is reasonable. But imagine he does, in fact, have such a duty, that his state believes that he has such a duty, and that it intends to discharge it on his behalf. It does so by, among other things, giving him the role at the {C}-node of paying a petrol tax. He begrudgingly pays his petrol tax, entirely out of fear of being punished if he does not. On our account, the sceptic's duty to pay his petrol tax cannot be characterized in the way that our simple story suggests. This is because he has not passed on his {A}-node duty: he has not willed, communicated the will, or taken steps towards reasonably believing that the state is pursuing that good on his behalf. (In fact, he may have taken steps to ensure the opposite.) Rather, the state has merely dissolved his {A}-node duty; the state secures the good and so there is no longer a case for the sceptic, at the {A}-node, to pursue that particular good. Our account is thus able to explain what goes wrong with the climate change sceptic's relation to his duty to pursue climate change mitigation: he has not discharged his {A}-node duty.

A second epistemically interesting case is one where the climate change sceptic does not believe there is a duty to pursue the good of climate change mitigation—and (let's imagine) he is correct—yet he does believe that the state intends to discharge that duty on his behalf. The fact that individuals often protest against actions taken by their states, using rhetoric like 'not in my name!', is at least anecdotal evidence for constituents having a general belief that when the state acts, it acts 'for' them or 'in their name', even when they do not endorse the state's actions and even when the state is wrong about their duties.[10] In these cases, the state intends to act on the individuals' behalf, and maybe the state believes it has done so successfully. But it fails to actually do so, since the individuals have not engaged in the transfer at {A}. They do not reasonably believe that the state has taken up their duties.

Of course, there are some goods that states have a duty to pursue, where individuals would not have duties to pursue those goods in some hypothetical 'state of nature'. Nonetheless, these duties of states are derived from individuals' {A}-node duties. Take the examples we started with: Australia's duty to legalize homosexual marriage; Britain's duty to spend less public money on the royal family; the United States' duty to make healthcare affordable. In the state of nature, individuals could not have had duties to pursue these goods. Without the relevant states having the goals and procedures they do, there would be no marriage institution in Australia, no British royal family, and no healthcare program in the United States. However, although Australia created its marriage institution, the duties to pursue the good of legalized homosexual marriage—given that there is a marriage institution—are {A}-node duties. Given the institution, it is plausible that individuals have a duty to pursue the legalization of homosexual marriage, which they can transfer to the state. It is a mistake to think about the transfer process in a state-of-nature way—to imagine the transfer from {A} to {B} all happening at some point in hypothetical history. The nature of the agent at {B} can generate new {A}-node duties.

The simple story can help us to understand, and to debate about, why states have certain duties that they have. On the simple story, they have them because the individuals at the {A}-node had duties to pursue some good. For any given duty, we can then have a debate about who these {A}-node individuals are, what their {A}-node duties are, and what should follow from the transfer. This gives us a starting point for thinking about, and refining, our account of particular duties of states. It gives us a useful way of starting discussions about the basis, and so about the implications, of states' duties.

---

[10] See e.g. US citizens' protests against the 'War on Terror' at <http://www.notinourname.net/> (accessed 1/12/2012).

Exceptions, complications, and clarifications for states

Two important exceptions to the simple story's application to states come from noticing that there can be partial variants of the thread from {A} to {C}. First of all, states can incur duties by making contracts with other agents (individual or collective). In these cases, the state's act of contracting will give rise to a collective duty that is justified entirely with reference to actions of the state itself. The state's {B}-node duty is not explained by any individual's {A}-node duties, to carry out the terms of the contract.[11] For some state contracts, for example, treaties regarding human rights, it is plausible that states' constituents do have relevant duties at the {A}-node. It is plausible that individuals have duties to be conditionally willing to play a part in respecting human rights. These duties can, we think, fairly be characterized as discharged by being passed on to the state, which acquires a duty to make contracts with other states regarding human rights. So the duty of states to make the contract derives from individuals' {A}-node duties. But even in these cases, the fact of the contracting will give rise to a new collective duty (a duty to abide by this contract, not merely a duty to form contracts regarding human rights) that is justified entirely with reference to agents at the {B}-node, that is, the state.

Second, the state itself can form and carry out a decision that causes harm. On most moral theories, agents have duties to remedy harms that they have done. So assume that the state, represented in the simple story at the {B}-node, would have duties to remedy the harm. Because collectives act by distributing roles to constituents, this will have implications at the {C}-node. The obligation to remedy historical injustice can be understood within this framework. But again, while this is part of the series of duties that we're interested in, it isn't a paradigm case because it doesn't include the duties at {A}. Any story about individuals at {A} transferring duties to the state at {B} will entail constituents' duties at {C}, because {B} and {C} are logically connected. But not all stories about collectives' duties at {B} will entail an explanation at {A}, because states are collective agents which can act autonomously, for example, in entering contracts and in undertaking potentially harmful actions, and they can thereby come to

---

[11] For example, Britain made an agreement with the International Olympic Committee that it would host the 2012 Olympic Games. In virtue of making this agreement, Britain incurred a duty to do what it could to see to it that it would host the Games. But it is implausible to think that individuals passed on duties to be conditionally willing to do what they could to see the Games hosted by Britain. They might have *chosen* to do what they could, but they had no *duty* to do so. And they might each have these duties *after* Britain has committed, but these duties arise at the {C}-node, not the {A}-node. But Britain's initial {B}-node duty to host the Games has no {A}-node *explanans*.

have duties to honour their contracts, make reparation for their harms, or perhaps even assist in virtue of their mere capacity to do so (in cases where transfer has not occurred).

In addition to these two exceptions in the case of states, there are complications that apply to the simple story, and equally to its application to states. First, just like other kinds of collectives, states can 'contract out' their duties (fulfilling duties by having another agent perform them, not to be confused with accruing duties by making agreements with other agents, discussed above). For example, a state might hire a private military company to assist it in waging a just war, or hire a national of another state to vote or argue on its behalf at a meeting of the United Nations. The firm or the delegate would thereby incur a duty, and the state would discharge its duty relating to the war or the vote by making the contract with the non-constituent and fulfilling its end of that contract.

Similarly, states might dissolve the duties of non-constituents, both collective and individual, so that some individuals start out with a duty but do not end up with one. For example, if New Zealand suddenly decided to take in all climate refugees from Pacific Island nations, then the duties of other states to take in these refugees would be dissolved. Those states' constituents would also have their {A}-node duties related to Pacific climate refugees dissolved. (This does not necessarily mean other states would be left with no duties. They might have a duty to compensate New Zealand. But the other states, and their constituents, would lose their duties to do what they can to take on the Pacific Islands' climate refugees.) Any individual who has her duties dissolved rather than transferred is not part of the transfer from {A} through {C}, although this should not be understood to imply that {A} and {C} must be symmetrical in composition. Asymmetries will sometimes be present in the simple story but will almost certainly exist when the {B}-node collective is a state, because constituents may transfer duties at {A} and then emigrate, or fall very seriously ill, or die, before they are able to be distributed duties at {C}.

Furthermore, a state cannot simply contract out whichever of its duties it likes. That is to say, just as there are agent-relative duties for individuals, such as duties of friendship (it matters that I comfort my friend rather than that just anyone comforts her), there are agent-relative duties for states. Consider the duty of Australia to apologize to Aborigines for the state's policy of forcibly removing Aboriginal children from their families. Here, Australia's duty is to see to that Australia apologizes to Aborigines, not just that Aborigines receive an apology. The duty cannot be contracted out; it must be done by a constituent of the Australian state (in this case, a suitably high-ranking constituent), acting within and because of their role, under instruction by the state itself, as a result of the state's decision-making procedure. When the constituent does this, Australia itself apologizes.[12]

Another seeming complication for states is actually no complication at all, but it is instructive to consider why. This is that it looks like certain duties, transferred from {A} to {B}, can be discharged at {B} without being distributed to {C}. Imagine that individual Canadians have a duty to do what they can to protect the human rights of those outside their borders, and they discharge this duty by transferring it to the state. The state can discharge the duty it thereby incurs by arguing a certain way in the next meeting of the United Nations, when human-rights protection is on the agenda. It looks like there is a transfer of duties between {A} and {B}, from individuals to the state, but that the collective duty can be discharged at {B} without needing to be distributed to constituents at {C}. However, that a collective duty can be discharged easily or by one person does not mean that it has been discharged without distribution of the necessary roles to constituents. Collectives act through their constituents. So the action of one constituent—who is the state's delegate to the relevant United Nations meeting and argues in a certain way—is the action of the collective. (If the state organizes for a non-constituent to act on its behalf, then this organizing action is an action of the collective, performed by a role-bearing individual at the {C}-node.) So rather than demonstrating that there can be a transfer of duties from {A} to {B} without any implications for {C}, this case demonstrates that the duties of constituents at {C} can be extremely minimal.

4. Implications for States

Failure at multiple stages

Understanding certain of states' duties via the simple story gives us the resources to criticize moral failure at multiple stages. Many of these varieties of state failure have an epistemic dimension. By noticing this, we can see the close connections between groups' epistemic lives and their moral lives.

First, states may fail to discharge their duties because the state has a faulty set of beliefs, desires, and goals, a faulty procedure for identifying future beliefs, desires, and goals, or a faulty procedure for distributing roles (*mutatis mutandis* for alternative conditions for collective agency). We can then locate the failure at the {B}-node, or perhaps with the individuals who set up the procedure, or (if

---

[12] As happened on 13 February 2008. See 'Kevin Rudd's Sorry Speech', *Sydney Morning Herald*, 13 February. 2008, online at <http://www.smh.com.au/articles/2008/02/13/1202760379056.html?page=fullpage> accessed 31/07/12.

they could have changed it) with the individuals at the {A}-node who did not make their state such that it could do their duty. Or perhaps the procedure and outputs of the state are fine, but a constituent has failed to perform her role—a failure which would be particularly egregious in a context in which all roles were necessary for the collective action's being performed. Here we can locate the failure at the {C}-node. Finally, maybe the {A}-node individuals failed to will that the state take on their duty, or to communicate this to the state, or to do anything to bring about their reasonable belief that the state would pursue the good on their behalf. Collectives can only do what individuals bring it about that they do, so here the failure would occur at the {A}-node. These different points at which the series of duties can fail are likely to have interesting implications for blaming states and their constituents: blame should be distributed differently, depending where in the series the failure occurs.

Take Australia's 2012 policy on refugees, for example: the 'Pacific Solution'. The policy, abandoned in 2007 and reintroduced in 2012, causes all asylum-seekers arriving by boat in Australia to be processed offshore in Nauru, and removes limits on processing time (as a deterrent to prospective asylum-seekers).[13] Assuming that such policy is unjust towards asylum-seekers, we can articulate the causes of this injustice by looking at each of the nodes in the simple story, applied to states.[14] Let us assume there is a valuable good which justifies duties for capable agents, namely that political asylum be provided to those who desperately need it.

At the {A}-node we have individual Australians, who might be failing in one (or more) of several ways. They might be failing to discharge their duties to pursue that good by failing to attempt to reform the collective of which they are a constituent. While individuals' duties at {A} can be discharged by forming, joining, reforming, or checking a collective, states are already a relatively fixed part of the political landscape, so individuals' duties will generally be to join, reform, or check. In the case of an unjust policy, it will most commonly be to reform. Or individuals might be failing to will that the collective takes up their duty. Or they might be failing to communicate their will to the Australian state.

---

[13] The bill supporting the policy passed on 16 August 2012. For more information, and an argument that the policy is unjust, see Bill Frelick, 'Australia: "Pacific Solution" Redux', *Human Rights Watch* online at <http://www.hrw.org/news/2012/08/17/australia-pacific-solution-redux> accessed 23/11/12.

[14] Not everyone would agree that this policy is unjust. However, we have intentionally selected an example in which 'ought implies can' allows individuals to have {A}-node duties that they can discharge by engaging with a state. Less controversial examples of unjust state policies are unlikely to meet this condition.

Perhaps they are not aware of their duties to pursue that good or just don't care about them.

At the {B}-node we have Australia, which has failed to take on the duties that Australians actually have (and has instead taken on duties it, or they, believe they have), and which distributes roles to constituents at {C} whose fulfilment will result in the state's pursuing morally objectionable ends. Perhaps Australia has been insufficiently responsive to individuals' attempts to transfer their duties to it. Or it is possible (though highly unrealistic) that it has taken on their duties, but has negligently or carelessly distributed roles in the wrong way.

And finally, at {C} we have the constituents of Australia, who may fail to perform their roles, which under some conditions will be sufficient to trigger no one performing their roles. In this particular example this may be a good thing—because it will mean Australia being thwarted in pursuing objectionable asylum policy. Yet it will be a bad thing whenever roles are distributed at {B} in accordance with the duties Australia actually has, which it in turn has in virtue of the transfer of duties individuals actually have at {A} because of some antecedently specified good.

Membership: robust duty transfer

A second implication of the simple story's application to states is that it can pick out a group of people who hold an interesting kind of relationship to the state: membership. 'Membership' is used in numerous ways by ordinary people to talk about their relationship with their state. It could pick out citizenship, permanent residence, national identity affiliation, and much else besides. The simple story can be used to develop one interesting way of specifying this complex notion, but it is by no means the only way.

The conception of membership that we will develop aims to pinpoint the relation between an individual and a state, where the individual's moral agency is bound up with the moral agency of the state, such that when we talk about whose moral agency the moral agency of that state is made up of in general, or distributes down to in general, we are talking about the members. A member is implicated in general in the state's fulfilment of or reneging upon its duties.[15] The members have a deep-seated moral-agential identification with the state. It is appropriate for these people to be ashamed or guilty when the state does wrong—even if, on this particular occasion, the member didn't intend for the state to pursue the given state of affairs on his behalf. As we shall see, the members have epistemic lives that are—and are likely to continue to be—intertwined with the

---

[15] By 'in general', we mean as opposed to 'for a specific duty'.

epistemic lives of their state: they and their state regularly transfer knowledge to one another, have various reasonable beliefs about one another's duties, and reliably recognize one another.

'Citizenship' is insufficient for picking out this relation, since in many states—for example, autocracies—citizens' agency is not implicated when the state acts. Their agency is not intertwined with that of their state. Another option is to say a person is a member if he engages in the simple story with a state on at least one occasion. This requires him to be a constituent, meeting whatever are the correct conditions for being part of a collective agent. However, again, he might be a constituent, and engage in the transfer process once, despite not being implicated in the state's agency in general. For example, a tourist might transfer to the state her duty to pursue systematized conduct on the road. That is, she might will that the state pursue that good in her behalf, communicate this by (say) tacitly agreeing to obey the road rules, and bring about her reasonable belief that the state pursues that good at least partly on her behalf. Moreover, it might be true that it does this partly on her behalf, and the state might give her a role related to this good—the role of obeying the state's road rules. So anyone who comes into the state's jurisdiction might be a constituent in the weak sense we outlined earlier, and might engage in the simple story (but only if the state actually pursues some good on their behalf). Because of this, is the tourist therefore a member of the state, in the sense that she is generally implicated in the state's agency? Common sense says not. If we allow individuals to be members just by happening to engage in this process with a state on one occasion, then we end up saying many people are members that common sense says are not members. We think that's worth avoiding.

We therefore suggest that the members of a state are the reliable or stable participants in the simple story where the collective at {B} is the state. By 'stable participants' we mean those who meet three conditions. First, a sufficient proportion of the transferred duties that the state has had during these individuals' lives has come from each of these individuals. (Exactly how many duties a state must have received from an individual for her to count as a member will depend on the capacities of the state, and on the duties (and outlets for collective transferral) that are open to that individual. Because these vary so greatly across both states and individuals, it is likely that there is no general answer to what a 'sufficient' proportion is.) Second, the state accepts the duty transfer and distributes roles back to each of these individuals, and has a history of doing so. Third, across a high proportion of likely futures of the actual world in which the state has a similar procedure to its current one (across which states are subjected to a wide range of differing economic, environmental, social, and cultural circumstances, both domestic and international), the state would accept the transfer of those individuals' duties, and distribute roles back to them. Membership, then, is both retrospective and prospective. In short, what makes members members is that: (1) they are now, and have in their past been, involved in every stage of the simple story for enough of the state's duties; and (2) this is likely to continue into the future, at least as long as the individual is alive and the state remains in more or less its current form.[16]

Who counts as a stable participant in the simple story with states will vary across time. Take the United States during the eighteenth century. Plausibly, at that time, no slaves would count as stable participants. This is because the United States was not engaging in the simple story with slaves' duties, nor had it been, nor would it if you had put (a similar version of) it in a range of economic, environmental, social, and cultural circumstances that were likely futures of that time. So slaves were not properly understood as partly constituting the state's agency at that time.

It will also vary in interesting ways at one time. The hermit in the woods, who wholly rejects the state—who does not will that the state acts on her behalf and does not appropriately use any roles she gets back—is not a member on our view, even if she is an enfranchised citizen. There is a significant sense in which she is not part of (does not partly constitute) the state. This is consistent with the citizen hermit's having a duty to be a member of the state, which she is failing to discharge. It is also consistent with her having a duty to obey the state's laws, which she might or might not be discharging. And it is consistent with the state's reliably dissolving the hermit's {A}-node duties (it might be much more efficient for the state to presume to act for everyone within the territory than to act for exactly those members who have transferred duties, and in so acting the state may dissolve the duties of non-members within the territory). Yet she is not stably transferring her {A}-node duties. Of course, one might think that this reliable duty-dissolving relationship is morally significant. If it is, then our account can agree: perhaps the citizen hermit has some morally significant relationship with her state. But this relationship is not one of membership, understood as partly constituting the state's agency.

On the other hand, consider a permanent resident alien—a non-citizen, for example, a New Zealander living permanently in Australia under the countries'

[16] It might also be that certain duties are more important than other duties for making someone a member: perhaps it's more important, for my being a member, that the state stably pursues on my behalf the good 'fighting all and only just wars', than that it stably pursues on my behalf the good 'breath-testing vehicle drivers'. We will not here take a stand on which duties, if any, might matter more for membership.

reciprocal immigration agreement—who pays taxes, votes, contributes to public debate, benefits from state-funded healthcare and education, and generally acts through the state on a range of issues—and does so stably. On our account, this person does count as a member. This seems the right result to us. She has, does, and probably will continue to transfer her duties to the state. Of course, she and the state might both say that she is not a member. Our suspicion is that, in saying that, they are using 'member' to pick out a different phenomenon than the one we are interested in. Perhaps they are using it to mean 'citizen' or 'national identity affiliate'. She is not a member in these senses, but her agency is implicated when the state acts. Many of the state's duties have, do, and probably will receive their justification from her (among others).

It is curious that members who are not legally recognized can achieve the duty transfer. In the case of the New Zealander in Australia, we have stipulated that she votes and pays taxes—so she has some level of formal, legal recognition. This makes her case somewhat easy. But what about permanent residents in countries that deny non-citizens these privileges? Can they realistically be members? Yes—but it will probably require some clever parasitism on others' legally recognized membership. When there are enough individuals who have a legally recognized status as a stable transferor, the modes of duty transfer are often quite open. Consider legal rights of protest, or petitioning, or writing to parliamentarians. These rights (and the associated formal processes) are no doubt designed for legally recognized members. However, others are more than capable of partaking in them, and thus regularly engaging in the {A} to {B} transfer. By obeying the laws that result, they also play a part in {C}. These individuals engage in an interesting kind of reverse free-riding: rather than selfishly exploiting the relationships that already exist, they add value by 'tagging along with' the official members.[17] These members are not as easy to pick out as citizens, but their relationship with the state is a valuable one. The simple story and our account of membership can explain what that value consists in: they are stable duty transferors.

It won't always be easy to identify the stable participants where the collective at {B} is a real state. We think that for democratic states, political participation is a good heuristic for stable transferral, and enfranchisement is a good heuristic for political participation. It's not perfect: the enfranchised hermit is not a member, while the disenfranchised but politically active permanent resident is. But when trying to identify members in the messy world of politics, it's a good enough place

---

[17] We are grateful to Miranda Fricker for discussion on this point.

to start. (A different story needs to be told for non-democracies, in which voters' decisions do not partly constitute the collective's decisions—because there are no voters. For dictatorships, for example, it is probable that the members will be only a small circle of elites.)

One might disagree with our intuitions about, for example, the citizen hermit and the politically active permanent resident. Perhaps one thinks that the hermit does partly constitute the state's agency, while the active permanent resident does not. There are a few responses to this, which are compatible with one another. First, we accept that our conceptualization might not capture all intuitions about membership. But it does capture a good number of them, and perhaps this is all that is needed to earn a theory serious consideration.

Second, perhaps some common-sense intuitions that our conceptualization does not capture should be rejected, in favour of conforming to a unified and well-motivated theory. In particular on the citizen hermit and the resident alien, we find it compelling that there must be something the citizen hermit can do to distance herself from the state. Is it really enough for membership simply that the state claims, has claimed, and might continue to claim that it represents her? We think not: the hermit's will must be involved in some way—plausibly, in the way we have described as being involved in discharging an {A}-node duty: she must do what she can to make the state capable of pursuing a good on her behalf, she must will that it do so, she must communicate this, and she must induce in herself a reasonable belief that the state is acting in accordance with that will. (In emphaszing individuals' intentions, our account has affinities with Pasternak 2013.)

Third, we can agree that some common-sense intuitions with which our account disagrees are picking up on a normatively significant relationship. But we can deny that that relationship is membership, understood as having one's moral agency bound up with that of a collective. Different theoretical conceptions of membership, and different intuitions, might be useful for capturing different phenomena, and maybe intuitions are not always clear on which phenomena they capture. To the extent that this last point is true, our proposed understanding of membership may not be a competitor to extant accounts in the literature, but may rather pick out something slightly different. (For other accounts of membership, see Walzer 1983; Parrish 2009; Stilz 2011; Pasternak 2013. Of these, Stilz's broadly Kantian account—on which individuals have duties to leave the state of nature and form states in order to secure the conditions necessary for justice—has the most affinity with our account. It would take at least another chapter to argue that our account is preferable to these. We present our account of membership merely as one possible account.)

5. Conclusion

Our starting point in this chapter was a simple story about the transfer of duties from individuals to collectives and back again. On that story, individuals can discharge their duties to do what they can (at not disproportionate cost) to pursue some good, by transferring that duty to a collective. The collective thereby acquires a new duty to do what it can (at not disproportionate cost) to pursue that good, which it discharges by distributing roles to constituents. Constituents thereby acquire duties to use their role with a view to pursuing the good in question.[18]

Our analysis depended on epistemic elements, including reciprocal recognition (members' recognition of their state, in willing it to discharge duties on their behalf, and the state's recognition of its members, in reliably distributing duties to them), intentional participation (of members by taking up the roles distributed to them by the state), bi-directional transfers of knowledge (members' communicating to their states that they wish them to act on their behalf, and states communicating to members the roles necessary for the securing of the goods they act in pursuit of), and, finally, degree of justification for beliefs (members being required to reasonably believe that their states will act on their behalf in taking up their duties).

Applying that simple story to individuals and states helps to shed light on the nature and justification of certain of states' duties, and on the possible sources of state failure.[19] An interesting upshot of this application is a new conception of who the members of states are, for the purpose of figuring out who, in general, is morally implicated in the state's actions. Our suggestion was that a state's members are those whose moral lives—and, therefore, whose epistemic lives—are robustly intertwined with that of their state.

---

[18] One issue that we haven't touched on is what the constituent's duty is when his role is pointless. It might be pointless either because the state's duty will not be discharged even with his role (since other constituents will fail to perform their necessary roles), or because the state's duty will be discharged even without his role (since other constituents will perform roles sufficient for discharging it). Lawford-Smith (2012) argues that collectives' constituents do not have duties to perform their roles where they have a reasonable belief that doing so would be pointless. We think this point applies to states' constituents as well.

[19] States are just one collective to which the simple story might be applied. It would be interesting to consider whether the simple story works in other contexts. Perhaps, for example, one could treat states as the agents at the {A}-node, and intergovernmental organizations (such as the North Atlantic Treaty Organization) as the agents at the {B}-node—thus ratcheting the whole story up one level, to groups in super-group contexts.

9

Four Types of Moral Wriggle Room

Uncovering Mechanisms of Racial Discrimination

Kai Spiekermann

1. Introduction

What we ought to do depends on facts. These facts can either be facts about the available actions and their context or facts about the applicable norms. Call the former 'action-facts' and the latter 'norm-facts'. If we are uncertain about these facts, we sometimes have the opportunity to acquire or ignore available information about them before acting. These opportunities can create strategic incentives. For example, you might think that you ought to comply with a norm that prohibits causing unnecessary greenhouse-gas emissions. You also have an inkling that your excessive air travel might be relevant in that context. However, you tend to avoid information about the (very high) levels of emissions caused by your flying, but you like to hear about how planes become more efficient, how other emission sources are much worse, and so on. In other words, you shape your own belief system about action-facts so that you can convince yourself that your behaviour is—by and large—compliant with the norm.

In a similar fashion, you can also choose to manipulate your beliefs about norm-facts. For instance, the norm prohibiting unnecessary greenhouse-gas

---

I am grateful to Miranda Fricker and Michael Brady for helpful comments. I gave this chapter at the Social and Political Thought Seminar, University of Sussex and the Departmental Seminar of the Department of Government, University of Essex, and would like to thank my audiences for their questions and comments.

emissions might not be uncontested. If so, you might be able to convince yourself that a more lenient norm applies, or that other people around you do not comply with the more stringent requirements anyhow, which, you might tell yourself, lowers the normative demand on you. If you succeed, you have strategically changed your beliefs about norm-facts to reduce your obligations. In both cases you have used 'moral wriggle room' (a term coined by Dana, Weber, and Kuang 2007)[1] to avoid costs of compliance.

There are two long and involved debates in epistemology that bear some resemblance to the issues addressed here, but are in fact conceptually distinct: first, the debate between doxastic voluntarists and non-voluntarists; second, the debate about the correct 'ethics of belief'. The first debate is about whether one can decide to believe something, or whether believing is an involuntary act (see, for instance, the contributions by Audi, Feldman, and Ginet in Steup 2001). Depending on one's position regarding the first debate, one is then likely to give different answers regarding the second debate, arriving at different views as to what we ought to believe or how we ought to form beliefs (cf. Kornblith 1983; Feldman 2000; Chignell 2013). Contributing to these debates is not within the scope of this chapter. All that matters for my discussion is the (much less contested) obligation to seek additional evidence if acquiring this evidence is easy and increases the chance of bringing about valuable outcomes. Or more precisely: the evidence is such that it enables a reasonable and well-meaning agent to perform actions with a higher expected (moral) value. In this uncontroversial sense, I maintain that there are *obligations of inquiry*; obligations to find out certain facts (with important normative implications) if there is a reasonable opportunity to do so (Hall and Johnson 1998). These obligations of inquiry are not primarily justified because they promote true beliefs or because of a general duty to ground our beliefs on evidence (as some evidentialists would have it); they are primarily justified by prudential and moral (not epistemic) reasons—the true beliefs promote valuable outcomes (Feldman 2000, 689).

Recent experiments in behavioral economics reveal that individuals frequently use moral wriggle room and often violate some fairly obvious obligations of inquiry. For example, several experiments show that 'strategic ignorance' can be reproduced in the artificial setting of a behavioural experiment. When people are put in situations in which obligations of inquiry apply, but complying with these obligations is potentially costly, a significant number of individuals choose to stay ignorant. Put in more psychological terms: individuals are subject to self-serving biases when deciding whether they obtain information that will have normative

---

[1] Though Dana, Weber, and Kuang prefer the term 'wiggle room' here.

implications. They tend to avoid information that promises to increase and seek information that promises to reduce the normative demands on them. In this sense individuals are engaged in *strategic normative context shaping*.

The experimental literature I refer to below tends to be focused on individuals because the empirical investigation of biases and systematic errors in human decision-making took individual behaviour as a starting-point. So what can we learn from this literature about 'The Epistemic Life of Groups'? How individual biases might scale up to the group level is an extremely important question, but one that is far from being settled, either theoretically or empirically. Sunstein and Hastie (2015, 52) suggest that group decisions often (though not always) amplify individual errors and biases. This is partly because individuals take the beliefs of others as evidence about what is true, and partly because conformity pressures stifle disagreement. The phenomenon of interest for this chapter, the strategic shaping of normative context, is also achieved more easily in a group because the individual strategic manipulation of information is often supported by collective biases and ignorance. For example, one's individual ability to maintain biased beliefs is much higher when all of one's peers have the same biased beliefs. The upshot is that while the current empirical evidence is largely about individual decision-making, it is at least suggestive of possible group biases, and plausibly these biases play out even more strongly in a group setting. In addition, possible measures to mitigate these biases are likely to be of a social nature: insofar as failures to meet obligations of inquiry are rooted in social practices and institutions, the problem must be addressed from a social perspective, the perspective of social epistemology.

In the following chapter I will report the results of some experiments that create a temptation for strategic information manipulation in different ways. I will then classify the results by developing a typology of 'moral wriggle rooms', before discussing implications for social moral epistemology, using the case study of racial discrimination.

2. Some Experiments

Strategic ignorance

Arguably the most influential paper on strategic ignorance in experimental economics is Dana, Weber, and Kuang's 'Exploiting Moral Wiggle Room' (2007). In their clever experimental design, subjects play a binary version of the so-called 'dictator game'. The baseline treatment is the game shown on the left-hand side of Figure 1. The 'dictator' (a more neutral term is used in the instructions for the

		S_1		States		S_2	
		payoffs				payoffs	
		Dictator	Receiver			Dictator	Receiver
Actions	A	6	1		A	6	5
	B	5	5		B	5	1

Fig. 1. Dana, Weber, and Kuang's binary dictator games (Numbers are payoffs in $US)

participants) can choose between actions A and B, causing payoffs for himself and his randomly matched 'receiver', as stated in the table. A selfish dictator will choose action A, as this maximizes his own payoff. A more altruistically minded dictator will sacrifice $1 to increase the payoff for the receiver by $4. In line with previous results from dictator game treatments, 74 per cent of dictators choose the 'fair' action B. In other words: considerations of equity often prevail in dictator games, even though the participants in the experiment will never learn with whom they played, and the game is played only once.

The clever part of the experiment is the 'hidden information' treatment. In this treatment, the dictators are told that the payoff table can either be the one in state S_1 (left) or S_2 (right) with equal probability. Note that the dictator payoff does not depend on the state; the uncertainty is only about the effect of actions A and B on *the receiver*. Before the dictators choose their action, they can optionally find out which payoff table applies by clicking on a button. Revealing the information is free and is literally no more than 'one click away'. Remarkably, now that dictators have a chance to remain ignorant about the effect of their action on the receiver, 'fair' choices become much less frequent: 44 per cent of dictators choose not to reveal which payoff table is applied. Of those, 86 per cent choose option A, thereby accepting a 50 per cent chance of only giving $1 to the receiver. By contrast, among those who reveal and find that S_1 is applicable (the right choice for S_2 is obvious since A is the better choice for both), only 25 per cent choose the selfish action A. The results show that even a minimal barrier to full information can reduce altruistic choices dramatically. Dana, Weber, and Kuang submit that these results suggest 'an illusory preference for fairness', that is, a willingness to show fair behaviour when the situation is clear, but to use ignorance as an excuse for selfishness when that excuse is available.

These results have had considerable impact among behavioural economists because they falsify many popular theories of altruistic behaviour. For example, if the dictator behaviour were driven by preferences over distributions alone (as is often assumed), we could not explain this dramatic reversal at all. Preferences-over-distributions theories have to be rejected because a dictator following such preferences would always reveal the information to satisfy his preferences. More promising are theories that appeal to self- and social image management (see Bénabou and Tirole 2011; Grossman and van der Weele forthcoming). Another approach is to think more explicitly about the prescriptions of the applicable norms that dictators feel subjected to. Spiekermann and Weiss (forthcoming) suggest that some subjects take the applicable norms as only prescribing altruism under certainty, while uncertainty renders more selfish choices morally acceptable. If that is so, the subjects do not have 'an illusory preference for fairness'. Rather, they think that the requirements of fairness under uncertainty differ from those under certainty.

Biased norm perceptions

Apart from strategically choosing (not) to know about action-facts, one can also acquire self-serving norm-facts. If one anticipates a decision that might come with significant norm compliance costs, it is tempting to believe that the most lenient norm is applicable. If there are several competing norms that could be applied, or if there are more or less demanding interpretations of a norm, biased beliefs about norm-facts are to be expected.

Devising experiments to measure strategic norm-fact acquisition is challenging: one needs to design a situation of normative ambivalence, create a choice situation in which the potential norms become relevant, show that the self-serving behavioural patterns arise, and—crucially—also show that these behavioural patterns are indeed caused by self-servingly chosen norms, rather than mere selfish payoff maximization. One influential experimental design to this effect can be found in Konow (2000); another relevant article is Bicchieri and Chavez (2013). Here I only report those parts of the experiments that are most relevant for biased norm perception and omit some of the more intricate details.

In the first stage of Konow's experiment, all subjects are asked to perform what experimental economists call a 'real effort task': they have to fold letters and stuff them into envelopes. In the treatments of interest here, the subjects are provided with material for ten letters and have seven minutes to complete them, which is more than enough for all of them to finish the task. Dictator–receiver pairs are formed randomly. To create a normatively ambivalent situation, Konow introduces arbitrary levels of credit (salary) for each letter completed, ranging from 55 to 75 cents for dictators and 25 to 45 cents for receivers, but averaging to 50 cents for every dictator–receiver pair. All subjects know the credits they and their counterpart receive. The total credits a pair has accrued can then be distributed between them by the dictator.

The subjects are told clearly that the differential credit for their work is entirely arbitrary, not related to their performance at all. This suggests that their joined credit ought to be distributed according to *output*, which is equal (both have completed ten letters). An alternative fairness norm, however, is to distribute according to *monetary contribution*, which varies because of the (arbitrary) difference in credits per envelope. Since the setup ensures that dictators always contribute more in terms of money, the latter norm[2] is attractive to them.

In a first step, the dictators are asked to distribute the credits between themselves and the receiver. Dictators keep, on average, 59 per cent of the money to be distributed, deviating statistically significantly from an equal distribution. This is probably due to selfish preferences.[3] In a second, unannounced round, Konow puts the dictators in the position of a benevolent dictator who has to distribute money between two other subjects (which means that the dictator has no economic stake in this decision). Ingeniously, these two other subjects are exact mirror images of the dictator and his receiver from round 1: they are assigned exactly the same credits for their envelope stuffing. In other words, the two subjects in round 2 exhibit exactly the same normative status as the dictator and his receiver in round 1. Asking the dictator to distribute as a benevolent dictator in this mirror image of the previous allocation decision enables Konow to observe *what the dictator perceives as fair if she is not personally involved*. This choice will, of course, be influenced by the previous decision and the potentially selfish choices made there.

In the second round, dictators still assign, on average, more to the person with higher per-envelope credit (their mirror image), even though the credits are known to be arbitrary, and even though the dictator has no economic stakes in this distribution decision (ruling out selfishness). This shows that some dictators have managed to convince themselves that subjects with higher per-envelope credits (like themselves) are more deserving. More precisely, after excluding dictators with theory-inconsistent choices (i.e. those keeping less than 50 per cent, or keeping less in the first than in the second round), 39 per cent show this bias. The choices can then be compared with a control group in which subjects decide only as benevolent dictators, without a prior round 1. In the control group, giving is completely in line with an egalitarian norm, so that the choices in the

---

[2] To be precise, Konow does not think that this behaviour can be considered norm-guided; for him the arbitrary credits offer a mere 'contextual pretense' (Konow 2000, 1079) for selfishness.

[3] In fact, in standard dictator games subjects tend to be more selfish and an average giving of 41% is quite high.

second round of the treatment group can only be explained by a self-serving norm bias.[4]

In Konow's experiments, the subjects convince themselves to apply a norm that works in their favour. In Bicchieri and Chavez's (2013) experiment, the subjects convince themselves that the norm in their favour is endorsed by others, and is therefore the relevant norm to comply with. Subjects play a so-called *ultimatum game*, in which the proposer can split $10 between himself and a receiver. In contrast to the dictator games discussed above, the receiver in the ultimatum game is not totally passive: in ultimatum games, the proposer suggests a split of the money, then the receiver can either accept the split (which is then implemented as suggested) or reject the split, which leads to zero payoff for both. One can interpret the receiver's reject option as a 'veto'. If the receiver is a payoff-maximizer and the ultimatum game is played only once (ruling out any reciprocity or reputation effects), the rational choice is to accept any positive offer, which, in turn, should induce the proposer to offer the minimum amount above zero to the receiver. This is not what experimenters find in the lab: the offer rates are much higher and rejection of low offers is frequent (see Camerer 2003, ch. 2 for details).

Returning to Bicchieri's and Chavez's experiment, the available choices are to split the money as $5–$5, $8–$2, or to have a coin tossed to decide between the two aforementioned splits. It is common knowledge which actions are available and which is chosen. Unsurprisingly, most proposers and receivers agree that an equal split is fair, while $8–$2 is unfair. There is, however, a substantial disagreement between proposers and receivers as to whether tossing the coin is fair: 81 per cent of proposers but only 51 per cent of receivers consider the coin toss fair. In addition, proposers were asked to estimate how many receivers think that the coin toss is fair; the average estimate was 76 per cent, while receivers only estimated an acceptance rate of 46 per cent for the coin toss among themselves.

This shows that the proposers are biased towards beliefs that, if widely shared, would constitute a norm that permits choosing the coin toss, while the receivers are much less inclined to believe that such a permissive norm exists. Since the coin toss leads to a higher expected payoff for the proposer, the results show a selfishly biased perception of norm-facts.

---

[4] A skeptic could could argue that the result is due to a minimal identification effect, such that the dictator identifies with his mirror image in the second round. This alternative explanation cannot be fully dismissed.

'Hiding behind small cakes'

The term 'hiding behind a small cake' was first introduced by the research group around Werner Güth in Güth, Huck, and Ockenfels (1996) and Güth and Huck (1997) in the context of ultimatum games. 'Hiding behind a small cake' is possible when the proposer either has a small or a large amount of money available for distribution. The receiver knows the probability for the existence of the small or large 'cake' amount, but does not know which 'cake' the proposer actually has available. Given that, the proposer can trick the receiver into thinking that a split is fair by offering 50 per cent *of the small cake amount*, even though the proposer has in fact received the large cake amount. This is indeed what some proposers appear to do (Güth, Huck, and Ockenfels 1996; see also Mitzkewitz and Nagel 1993 and Güth and Huck 1997).[5]

The obvious problem with using the ultimatum game is that we cannot distinguish the proposer's motivation to maintain a positive image from the payoff incentive to avoid an offer rejection. In other words, we cannot determine whether the proposer gives 50 per cent (or a little less) of the small cake for tactical reasons, or whether he chooses his offer because he likes to appear fair *regardless of the monetary payoff*. To investigate the latter motivation separately, one needs to remove the tactical incentive. This can be done by creating an opportunity to hide behind a small cake in a dictator game.

Ockenfels and Werner (2012) did precisely that. In cooperation with a large German newspaper, they had 701 members of the general public play a dictator game with two possible cake sizes, administered through the website of the daily *Die Welt*. The 'cake' was either 1,000 or 3,000 Euros. Participants know that one randomly selected receiver–dictator pair is paid out with real money. Ockenfels and Werner assign their subjects to one of two treatments. In NOINFO, the receivers only learn how much money the dictator assigns to them; no other information is provided. In INFO, the receivers are also told whether the dictator had the small or the large cake available (and the dictators know that the receivers will find out).[6] While dictators can hide behind a small cake in NOINFO, their cover is blown in INFO, so that one would expect a significant difference in

---

[5] Though most of the evidence is of an anecdotal nature—there is surprisingly little rigorous testing for 'hiding behind a small cake' in the ultimatum game.

[6] To obtain as many data points as possible, all subjects are asked what they would do as dictators for two cases: if the cake is large and if the cake is small. Roles and cake sizes were assigned subsequently, and the stated strategies were then implemented for the dictators.

giving between the two treatments at and just below 500 Euros (= 50 per cent of the small cake).

This is indeed what Ockenfels and Werner find—though the effect is perhaps not as strong as one would have thought: in the INFO treatment, 10.4 per cent of dictators choose to give 500 Euros or less when they have a large cake, compared to 14.8 per cent in the NOINFO treatment, a statistically significant difference. The effect is more pronounced when looking at those dictators giving *exactly* 500 Euro when they have a large cake (2.6 per cent versus 7.6 per cent). Since 500 out of 3,000 is not a particularly plausible fraction to give (unless one wants to hide behind a small cake), these results are quite revealing. They show that some subjects, given the opportunity, pretend to have a small cake when they really have a large one.

3. Four Kinds of 'Moral Wriggle Room'

The experiments described above show that self-serving information manipulation can work in different ways. It is useful to distinguish two dimensions. First, does the manipulation target information about action-facts or norm-facts? Second, is the primary target of manipulation one's own belief, or someone else's belief?

Beginning with the first dimension, it is useful to describe *norm-facts* more carefully: they are facts about the norms prescribing appropriate actions in a specified context. In ambivalent situations, individuals might disagree about the applicable norms, and this disagreement can be fueled by selfish biases. In addition, in the case of conventions and social norms, the existence of social practices or social expectations partly determines the normative content, since social norms and conventions hinge on what others do or want us to do (Bicchieri 2006; Southwood 2011). Individuals might therefore develop a conveniently biased perception as to what others do or expect, as in Bicchieri's and Chavez's experiment.

I take *action-facts* to be facts about the outcomes of actions and all other potentially normatively relevant facts, apart from the norm-facts. In the case of Dana, Weber, and Kuang's hidden information treatment, there is an obvious action-fact: the dictator either does or does not cause a negative externality for the receiver. Avoiding this information is potentially advantageous, as one can prevent the direct confrontation with a demanding norm that prohibits the imposition of such severe losses. In the case of hiding behind a small cake, the relevant action-fact is, most immediately, about the endowment of the dictator (the small or large cake).

In both of these experiments, the relevant avoided facts pertain to the outcome, described as the payoff distribution between dictator and receiver. However, action-facts are not necessarily payoff distributions, they can also be about other normatively relevant properties. For instance, in an experimental setup in Spiekermann and Weiss (forthcoming), dictators can decide (not) to learn about how deserving their receiver is. There, identical payoff distributions can be part of two quite different outcomes: giving, say, 10 per cent of the endowment to an undeserving receiver may be fair, but giving 10 per cent to a deserving receiver may be blatantly unfair. Here the relevant facts are about desert or entitlement, not payoff distributions.

The second dimension pertains to the person targeted for belief manipulation. The more obvious way is to manipulate one's own beliefs about the situation by being biased in a self-serving way. However, the 'hiding behind a small cake' literature shows that it is also possible to manipulate the beliefs of others in order to reduce expectations on oneself. By making partial or incomplete statements about the world, or by using ambiguity, subjects can create false impressions about normatively relevant facts.

Manipulating the norm-facts held by others is the least researched type of moral wriggle room. Nevertheless, I think it is more than a mere conceptual possibility. For instance, one could make self-serving statements to others about the norms that apply or make misleading suggestions about which norm is more widely accepted or which behaviour would be accepted by the relevant peer group. Such self-serving manipulations of others's beliefs might be particularly relevant when the target person is uncertain about or unfamiliar with the normative context. For instance, waiters or cab drivers might want to suggest overly generous local tipping norms to uninitiated tourists. To the best of my knowledge, an experiment to that effect has not yet been conducted.

Table 1 summarizes the two dimensions and the resulting four types of moral wriggle room and positions the three types of experiments from the previous section in the appropriate cells.

Table 1. Experiments and four types of moral wriggle room

Manipulate information about...	Target own beliefs	Target others's beliefs
action-facts	Strategic Ignorance	Hiding Behind a Small Cake
norm-facts	Biased Norm Perception	?

4. Wriggle Room, Social Moral Epistemology, and Racism

We have seen that the manipulation of information about action- and norm-facts, accessible to oneself or to others, is not only conceptually possible, but can be reproduced under the rigorous control conditions of a behavioural lab. But can we use the results in those stylized, artificial environments to learn something about self-serving tendencies in real-world knowledge acquisition? I suggest that we can. In this final section I draw links to the field of social moral epistemology. I also show that the more abstract experimental results are at least suggestive for understanding subtle forms of racism and other forms of discrimination.

Allen Buchanan (2002, 126) defines social moral epistemology as the 'study of the social practices and institutions that promote (or impede) the formation, preservation, and transmission of true beliefs so far as true beliefs facilitate right action or reduce the incidence of wrong action'. Moral wriggling is one important mechanism for explaining how social practices can impede the formation of true beliefs, and how the resulting biased beliefs lead to wrong actions. The self-serving biases identified above allow individuals and even whole societies to get away with acts that are objectively impermissible, sometimes without even realizing it themselves.

As Miranda Fricker observes, the field of epistemology has a tendency to emphasize ideal standards of knowledge acquisition, losing sight of epistemic injustice even though 'the only way to reveal what is involved in epistemic justice (indeed, even to see that there is such a thing as epistemic justice) is by looking at the negative space that is epistemic injustice' (Fricker 2007, viii). Understanding the strategies for avoiding normatively relevant knowledge will allow us to grasp more clearly how epistemic injustice can be the result of subtle biases. We will see that the problem of moral wriggling can lead to particularly serious injustice because it enables individuals to hide their failings to themselves or others, leading to entrenched, stealthy practices of injustice that need to be uncovered.

For a case study, consider the many subtle forms of racism. Sometimes we encounter transparently racist attitudes. An overt, conscious racist holds a certain set of beliefs, norms, and values, and these are typically clear to himself and others. Such racists are easy to recognize and argue against. But there are also much more subtle forms of racism—racially biased attitudes and actions that are less transparent, rooted in the moral wriggle room discussed above. José Medina refers to such subtle mechanisms when he describes how 'epistemic neglect' can lead to racially biased perception:

Continual epistemic neglect creates blinders that one allows to grow around one's epistemic perspective, constraining and slanting one's vantage point. As we shall see, responsible epistemic agency requires a minimum of diligence, because knowledge requires work and its acquisition will not happen without the active participation of the knower. Becoming lazy is letting oneself go epistemically; and it damages the objectivity of one's perspective and limits one's epistemic agency. (Medina 2013, 33–4)

Medina emphasizes that epistemic neglect is often unconscious, but that does not imply that it is harmless or exculpating:

Actively ignorant subjects are those who can be blamed not just for lacking particular pieces of knowledge, but also for having epistemic attitudes and habits that contribute to create and maintain bodies of ignorance. These subjects are at fault for their complicity (often unconscious and involuntary) with epistemic injustices that support and contribute to situations of oppression. (Medina 2013, 39)

Charles Mills[7] also emphasizes the possibly unconscious but still pervasive nature of this mechanism when he observes that:

racialized causality can give rise to what I am calling white ignorance, straightforwardly for a racist cognizer, but indirectly for a nonracist cognizer who may form mistaken beliefs (e.g., that after the abolition of slavery in the United States, blacks generally had opportunities equal to whites) because of the social suppression of the pertinent knowledge, though without prejudice himself. (Mills, 2007, 21)

Mills's 'white ignorance' is structurally similar to the strategic ignorance detected by Dana, Weber, and Kuang. Their experiment demonstrates that many subjects are keen to avoid information that would clarify and potentially increase the normative demands on them, and do so even if that information is free and the potential negative implications of ignorance clear. Since information avoidance works for these subjects in such an artificial setting, it should be even easier in real-life interactions, where we make plenty of choices to avoid information. Whether we read or not read a newspaper article, watch or not watch a TV programme, talk or not talk to an acquaintance—we take all these decisions on a daily basis, and often we have an inkling whether we can expect to learn something that might be 'uncomfortable' news for us. For instance, if we were to put away Medina's 'blinders' we might be confronted with the fact that there weren't equal opportunities for African Americans after the abolition of slavery, which would, in turn, entail obligations to address such a continuing injustice.

As recognized by Medina, strategic ignorance can work in an unconscious way. The more conscious the subject becomes of the implications of ignorance, the

---

[7] Thanks to Miranda Fricker for pointing me towards Medina and Mills.

more she may become aware of her 'epistemic laziness', and feel compelled to do something about it. However, such obligations of inquiry seem to have a comparatively weak effect on many subjects: if the effect were stronger, all subjects in the Dana, Weber, and Kuang experiment would have clicked the button to reveal information.

A different form of wriggle room and strategic belief manipulation suggested by the experimental literature (especially in Spiekermann and Weiss forthcoming) features less prominently in the literature on racism, but might also be important: the opportunity to selectively *acquire* the sort of information that justifies one's potentially problematic moral conduct. For instance, in many field experiments researchers found robust evidence that applicants with 'foreign-sounding' names on their written application materials are less likely to be invited to job interviews, despite having qualifications equal to the control group (e.g. Riach and Rich 2002, for a review). It is, of course, perfectly possible that the relevant selectors are conscious racists. Much more plausible, however, is the assumption that the racist bias is at least partially unconscious. The selectors might focus on the weaknesses (and overlook the strengths) of a perceived 'non-native' applicant more than they would otherwise, without noticing their own biased perception. They might therefore feel entirely justified in their decisions and not see any racial bias. If this is true, the problem is not rooted in an open racist attitude, but in more subtle biases that might go unnoticed.

So far the focus has been on the upper left cell of Table 1. However, other forms of moral wriggling when it comes to racism are also conceivable. Many people swimming along with the stream in a racist society reduce their cognitive dissonance by convincing themselves that norms of equal treatment apply only in certain contexts, or that they experience strong normative expectations by others that 'force' them to engage in racist practices, even though in reality the sanctions for non-compliance might be minimal or non-existent. Again, the choices here may be both to know and not to know about the relevant norms.

Finally, moral wriggling, as we have seen, might also involve manipulating the information others have. The artificial setting from 'hiding behind a small cake' is probably not directly transferable to the context of race. Nevertheless, it is not too difficult to think of mechanisms that skew the normatively relevant information others hold. For example, Medina reports that the state of Arizona decided to ban 'ethnic studies' in schools. This included removing several texts from the curriculum on the grounds that they promote 'critical race theory'. Medina interprets these interventions as an attempt to remove intellectual resources that would help to uncover racial discrimination. Banning these topics and texts prevents pupils from experiencing the necessary 'friction' that makes them question the pre-dominant 'cognitive laziness' (Medina 2013, 145). For a less direct example, consider the notion of 'colour-blindness' as a supposed ideal of equal treatment. Is it a fanciful assumption that 'colour-blindness' might have an epistemic-strategic function—namely, to stop others from questioning entrenched injustice? Being 'colour-blind' sounds like a normatively compelling idea on first sight, but it effectively blocks any discussion of existing racial discrimination if 'colour-blindness' is taken to rule out *any* reference to race in the public sphere. Such a blanket ban prevents citizens from uncovering normatively relevant facts and norms that would undermine the status quo and lead to more demanding normative-political obligations to tackle discrimination.[8]

As we have seen, 'moral wriggling' can often operate in an unconscious, yet systematic way. This chimes well with a recent turn of attention towards more subtle forms of racial biases. Elizabeth Anderson puts these biases at the centre of her analysis by emphasizing 'the roles of implicit and automatic cognition, which cause discriminatory treatment even in the absence of discriminatory beliefs or a conscious intention to discriminate' (Anderson 2010, 63). In a similar vein, Medina stresses the systematic and social nature of these biases:

We become active participants in collective bodies of ignorance typically without knowing it and apparently without much conscious effort on our part, but this is because there is a complex set of social structures, procedures, and practices that encourage us to go on with our daily business without taking an interest in certain things. (Medina 2013, 145)

As a final step in our analysis of moral wriggling as a problem of social epistemology, we can also ask which mechanisms and institutions can be used to mitigate this problem. A full analysis of counter-strategies is clearly beyond this short chapter (but see Anderson 2010 and Medina 2013 for book-length treatments of that question with regard to racism). Here a short discussion of some helpful, suggestive experimental results must suffice.

The strategic avoidance or acquisition of relevant information is possible because individuals do not comply with their obligations of inquiry to be informed enough in order to choose the right actions. How far these obligations of inquiry might go is subject to debate, but in the experiments presented and in the case of 'epistemic laziness' with regard to racial discrimination, it should be uncontroversial that individuals are under *some* obligation to understand the context they are acting in. The question is how one can prod individuals to meet this demand.

---

[8] See Anderson (2010) for a discussion as to how the principle of colour-blindness leads to additional discrimination outcomes under non-ideal circumstances.

An experiment by Cappelen *et al.* (2011) suggests that structured normative reflection can be an important tool. Cappelen *et al.* let their subjects play a dictator game with money earned in a previous investment task. Similar to Konow's (2000) setup, the subjects have different rates of return in the investment task, but the rate is assigned at random to create normative ambiguity. The profit in the investment stage is partly due to individual choices, partly due to luck. In the treatment group, the dictators have to reflect on the fairness of several distributive principles before playing the actual dictator game themselves. Cappelen *et al.* suggest that relevant fairness considerations could be 'to share equally, to share in proportion to individual investment, and to share in proportion to individual production' (p. 107). Without going into the details of their analysis, their experiment shows that both the average and the mode of giving increases with the reflection treatment, providing some evidence that more selfish choices are harder to make after reflecting on fairness principles. Increasing the salience of fairness principles (in particular, making the subject's own fairness principles salient) can sometimes completely eliminate self-serving biases, as Haisley and Weber (2010) show.

An experiment that is calibrated specifically to test measures against strategic avoidance or uptake of information is, unfortunately, still missing. Nonetheless, we can take a lesson from the experiments mentioned: if we hope for altruistic behaviour based on norms that are not enforced by heavy-handed monitoring and sanctions, then the effectiveness of these norms can be improved by either removing uncertainty about the situation, or, if that is not possible, by simply reminding individuals of their own principles. In the case of racism, this could be put into practice by making the experience and the effects of racism much more concrete in the public debate. For instance, many whites are aware that blacks tend to experience some disadvantages. What they are likely to underestimate (and perhaps ignore strategically), is the pervasiveness and strength of the discrimination. One could try to counter this with education programmes that make the experience of discrimination concrete in terms of the lived experience, but also in terms of differences in educational and job prospects, wealth, health care, life expectancy, and so on. The 'blinders' Medina talks about could be removed by making the ongoing, experienced discrimination salient and thus impossible to ignore.

Another option to counter epistemic biases is to strengthen the norms regulating our obligations of inquiry. So, rather than making people aware of the specific bads caused by their actions, one could train them to *be more aware of their obligations of inquiry*, in a general sense. To what extent this can succeed is an empirical question. At this point, little is known about possible effects of strengthening the norms of inquiry. *Ceteris paribus*, making specific normative properties of an action salient is probably easier than training people to identify these salient properties themselves—even though the latter would often be the more robust solution. Training individuals to resist self-serving biases would enhance their autonomy, but it remains unclear whether it can succeed.

5. Conclusion

Drawing on the experimental literature in behavioural economics, I have identified four different types of moral wriggle room and linked these types to some recent behavioural experiments. I have shown that the taxonomy and the experimental results reported can be applied to the problem of racial discrimination. Thinking about the effects of 'moral wriggling' in a social context reveals that the effects might be even stronger in real life and thus a graver cause for concern. The often unconscious nature of the self-serving biases investigated challenges us to think about collective counter-strategies to uncover and prevent the harm caused by 'moral wriggling'.

PART IV

Philosophy of Science

10

Collective Belief, Kuhn, and the String Theory Community

James Owen Weatherall and *Margaret Gilbert*

1. Introduction

Authoritative scientific assertions often take the form of an ascription of belief to a particular population of scientists. For instance, one regularly hears or reads statements such as, "physicists believe that elementary particles obey the laws of quantum mechanics," or "biologists think the chimpanzee and bonobo share a recent common ancestor." A chemist might say, speaking of his colleagues, "We believe that there may be additional undiscovered elements." Let us refer to these statements as assertions of *scientific consensus*.

Many philosophers take understanding the development of and (especially) changes in scientific consensus to be essential to understanding science. Despite their ubiquity, however, statements of the above form may appear puzzling. For it is natural to construe them as ascribing beliefs not to individual scientists, *seriatim*, but rather to collections or communities of scientists—to physicists, for instance, as a group. One may wonder what a *group's* belief can amount to.

Evidently, if one wants to understand the nature of scientific consensus, and by extension, change in scientific consensus, one would do well initially to explore

---

The authors are grateful to Jeff Barrett, Cailin O'Connor, and K. Brad Wray for helpful comments on a draft of this article, and to audiences at UC Irvine and Objectivity 2010, a conference held at University of British Columbia, to which a previous draft was presented. Thank you, too, to Michael Brady and Miranda Fricker for inviting us to contribute this article to this volume and for their helpful comments on a draft.

what is intended when one ascribes a belief to a group. Several proposals have been made on this score in the literature.

According to one immediate and widely accepted suggestion, a group of scientists, for instance, is rightly said to hold a given belief about a relevant scientific topic just in case all or nearly all members of the group holds the belief (cf. Quinton 1975/6, 17). And perhaps in some cases, this or something very like it *is* what is meant when one says that a group has a belief. But there is also a radically distinct possibility.

In her writings on the subject one of us (Gilbert) has pointed out that in standard cases of ordinary usage, when one ascribes a belief to a group, one seems not to be asserting that all or most of the members of the group hold the belief. Indeed, it often seems that group beliefs are not a matter of the beliefs of the members of the group at all—that is, it is neither necessary nor sufficient for a group to hold a belief that all or most of its members hold that belief. The account of group belief that Gilbert has developed over the last several decades (see e.g. Gilbert 1987, 1989, and 2002) accords with this observation.

It is now standard to call the first sort of account of group belief a "summative" account. Gilbert's, in contrast, is a "collective" account.

Gilbert's account was intended to capture a central intuitive meaning of assertions of the general form "G believes that p", where G is some group of people. On this account, members of groups that hold beliefs in this collective sense have certain obligations regarding the belief. And it is the ground of these obligations, and not necessarily the beliefs of the members of the group, that determines whether a group has a belief.[1]

Suppose now that assertions of scientific consensus are accurate ascriptions of group beliefs to scientific communities, in the collective sense of group belief described by Gilbert. In that case, members of a scientific community holding a consensus have the obligations associated with being party to a group belief.[2] In "Collective Belief and Scientific Change" (2000a), Gilbert suggests that this would

---

[1] The ground of the obligations is what Gilbert refers to as "joint commitment." See the next section.

[2] Wray (2001) and others have argued that one should not think of the phenomenon Gilbert describes as group "belief," but rather as group "acceptance," because the phenomenon differs in certain ways from belief as traditionally understood. We will not address the belief/acceptance worry here (for more on this, see Gilbert 2002 and Gilbert and Pilchman 2014). The important point for the present chapter will be that members of certain scientific communities act as though there is a *joint commitment* with a particular content in place. Whether this joint commitment is better referred to as constituting a case of belief or, rather, acceptance has no direct bearing; for convenience, we will continue to refer to the phenomenon in question as group belief.

have consequences for the conditions under which scientific change can occur. In particular, she argued that the obligations associated with a group belief on her account can act as barriers both to the introduction of new ideas by members of the group and to the fair evaluation of ideas proposed by experts outside the group.[3] This suggests a way of explaining some of historian of science Thomas Kuhn's best-known observations on scientific change, in particular his focus on the way "normal science" proceeds, a point we develop in what follows.[4]

This chapter focuses on some commentary, by contemporary physicist Lee Smolin, on certain "sociological" features of a particular high-profile branch of contemporary physics known as string theory, and the way in which his observations can be explained in Gilbertian terms, which are in turn supported by these observations. It thus presents a particular case-study in relation both to Kuhn's and Gilbert's work—with an emphasis on the latter.

The commentary we discuss arises in connection with Smolin's (2006) criticism of string theory. In particular, Smolin argues that string theory has failed to provide an adequate or compelling account of nature, and yet it continues to have a dominant role in theoretical physics (or at least, in high-energy physics, which is concerned with elementary particles). As a result, high-energy physics has failed to make significant progress in the last thirty years. In part, Smolin claims, the continued interest in string theory despite the problems he identifies can be explained by appealing to sociological factors. He goes on to list "seven unusual aspects of the string theory community" (p. 284) that, on his view, have contributed to the recent failure of high-energy theory.

For our purposes here, we take no stand on whether Smolin's broader critique has merit, or whether string theory is a fruitful or promising research program. Rather, we focus on the sociological considerations Smolin raises. Our purpose is to explore the relationship between the features one would expect of a group

---

[3] Much of the literature on collective epistemology has focused on group *knowledge*, rather than group *belief*. (For examples, see Longino 1990; Kitcher 1994; Knorr-Cetina 1999; Giere and Moffatt 2003; Mathiesen 2006; Wray 2007; Rolin 2008, 2010; and Fagan 2011, forthcoming *a*, *b*. See also Schmitt 1994.) This debate, however, is orthogonal to the present discussion, as the question we are addressing here is not whether a scientific community has knowledge, but rather whether certain features of one such community are well explained by the hypothesis that the community has a group belief, with its attendant obligations. Such group belief may, of course, be a central component in group knowledge, at least on a collective construal; that is a topic on which we do not attempt to pronounce here.

[4] Kuhn (1962). Gilbert had privately noted this connection without planning to write about it. The present discussion was prompted by Miranda Fricker, commenting on a draft of this chapter. For another brief discussion of Gilbert's ideas in relation to Kuhnian normal science, see Bird (2010).

holding a belief on Gilbert's account and the features of the string theory community as described by Smolin.[5]

We argue that the features that Smolin ascribes to the string theory community are precisely what one would expect if Gilbert's suggestion that such communities hold group beliefs on her account is correct.[6] In other words, *if* Smolin's picture of the sociology of string theory is accurate—something we assume for the sake of our discussion here—then the string theory community has properties that are very well explained on Gilbert's account of group belief, applied to the string theory community with the assumption that the community holds relevant group beliefs.

Gilbert's account offers, in particular, an effective explanation of why the string theory community may appear to act *irrationally* with regard to certain countervailing evidence.[7,8] It is here that the connection to Kuhn arises. We will suggest

---

[5] As Gilbert points out (2000a), whether and to what extent scientific communities hold group beliefs is ultimately an empirical question. In this regard, the present chapter can be understood as a study in the vein of Beatty (2006) and, especially, Bouvier (2004), which seek to evaluate whether real scientific communities are well described as groups with group beliefs by studying these communities.

[6] Wray (2007) has argued that *research teams* can have group beliefs (or rather, group acceptances—see n. 2), but that *communities of scientists associated with research fields* (such as the community of astrophysicists or of biologists) and *the community of scientists* as a whole cannot. (For a compelling reply, see Rolin 2008.) It is not clear where string theorists fit into Wray's categorization of groups of scientists—surely string theorists are not systematically organized as a research team, and yet they are only a sub-group of high-energy physicists. In any case, the present claim is that one does well to understand string theorists as the kind of community that not only *can*, but *does* have group beliefs. One may well take this as an argument against Wray's view that only research teams can have group beliefs. We shall not, however, attempt to engage with the specifics of Wray's discussion here.

[7] Wray (2001) and Mathiesen (2006) have argued that groups holding group beliefs on Gilbert's account may not be epistemically rational, in the sense that group beliefs may not be updated appropriately in light of new evidence. This is taken to be a criticism of Gilbert's view. We will not address this criticism in any detail in the present work, though we will note that if, in fact, groups holding group beliefs *do* fail to account for evidence in an epistemically responsible way, then it is a *virtue* of Gilbert's account that it appears to predict this behavior. (For another response to Wray and Mathiesen, see Rolin 2010.)

[8] Gilbert (2000a) argues that postulating that certain scientific communities hold group beliefs explains certain features of scientific communities and theory change related to resistance to heterodoxy. This argument has been criticized by a number of authors, including Wray (2006, 2007), Rolin (2008), and Fagan (2011), on the grounds that it is not clear how generic the features Gilbert mentions are, and thus that it is not clear that such features stand in need of explanation. For present purposes, one can set this general discussion aside and focus on the *specific* explanatory question presented by Smolin's work, namely, does Gilbert's account provide a compelling explanation of the striking features of the string theory community? If so, then it seems that there is a potentially interesting question to be explored concerning the role of group belief in at least some scientific communities.

that the apparent irrationality of the string theory community, as described by Smolin, is symptomatic of the conservatism Kuhn attributes to normal science.

At the end of the chapter we will address an apparent conflict that arises between our argument here and Smolin's characterization of his seven features, related to the fact that he explicitly characterizes the features as *unusual*, whereas if Gilbert's account is correct, one should expect such features to be exhibited by *any* group holding a belief—including any scientific community holding a consensus view. We will argue that these features seem unusual to Smolin not because they are actually unusual, but because he occupies an unusual position from which to observe them.

We should note that the goal of this chapter is not to evaluate or dispute Smolin's picture of the string theory community. Likewise, we will not directly evaluate Gilbert's account of group belief. Any evaluative content with regard to Gilbert will be implicit in how effectively her theory accounts for the string theory community, at least as described by Smolin. Similarly, insofar as Smolin's perceptions of the string theory community accord with Gilbert's collective account of group belief, which was arrived at independently by reflection on everyday thought, talk, and behavior in multiple domains, the latter has some tendency to confirm the accuracy of the former.

The remainder of this chapter is organized as follows. We begin with a brief presentation of Gilbert's theory of social groups and group belief, along with its connection to scientific consensus and change, as described in "Collective Belief and Scientific Change" (2000*a*). We then link her theories to Kuhn's observations on scientific change. Next, we give a brief history and description of string theory, explaining how it has now come to have a dominant role in the broader high-energy physics community. Finally, we discuss the seven points of Smolin's sociological critique in light of Gilbert's theory of group belief, arguing that the striking features of string theory Smolin describes are well explained by the hypotheses that group beliefs on Gilbert's account are at issue and that the string theory community holds relevant group beliefs.

2. Gilbert on Collective Belief and Scientific Change

Gilbert's account of group belief offers an interpretation of such statements as: "The seminar believes that the second article is less convincing than the first"; "The European Union believes that the Euro will rebound in light of the economic indicators released today"; or "Ben and AJ believe Jim is coming, even though he's already a few minutes late." In each of these sentences, a belief is ascribed to the subject. These subjects are not individual human beings, but rather something that is made up of human beings. One might call them "groups of people," or just "groups," to capture that they refer to something of which several (or many) people are members. The problem, then, is to understand what such "group belief" amounts to.

Gilbert's account of group belief can be expressed as follows.

A group G believes that *p* if and only if the members of G are *jointly committed* to believing that *p* as a body.[9]

When this condition obtains (and only then), each member of G may truly say "We believe that p," where "we" is intended with respect to G.

It is worth clarifying first that the phrase "as a body" is not meant to be sacrosanct. "As a unit" or "as one" would do just as well. The idea is that the object of the commitment, the thing being committed to, is to emulate as far as possible a single believer of the proposition in question, by virtue of the combined actions and utterances of the parties.

This analysis relies on the technical concept of "joint commitment." We will not define this term outright; instead, we will explain it by describing some central features of a joint commitment.[10] When people are jointly committed in some way, they are in a particular normative situation—each is committed to act in a certain way—as a result of a specific process that involves them all. As to that process, it is common knowledge[11] between the people in question that each of them has expressed his or her personal willingness together with the others to commit them all in the way in question. Thus, in the case of a group belief that *p*, each member of G has expressed his or her willingness together with the others to commit them all to emulating a single believer of *p* by virtue of their combined utterances and actions, and it is common knowledge among the members of G that this is the case.

It is important to emphasize that in order for the parties to be jointly committed, it is neither necessary nor sufficient for the members of G to each make what Gilbert refers to as a *personal* commitment to do his or her part in emulating, together with the others, a single believer of some proposition. A personal commitment in Gilbert's sense would be engaged by Jim's decision

---

[9] For more on Gilbert's account, see e. g. Gilbert (1987, 1989), her earliest discussions, and (2000a, 2004).

[10] For a longer but still compact general introduction to the notion, see Gilbert (2013), Introduction, and for further details, see ch. 2 of that work.

[11] "Common knowledge" is usually construed as a technical term in philosophy and the technical sense we have in mind is akin to Lewis (1969). However, for the current purposes the intuitive concept is sufficient. We will not later make any use of a specific formulation of common knowledge.

to work late tonight, for instance. A joint commitment is not composed of personal commitments, which are the unilateral product of the committed person's will and unilaterally rescindable by that person.

In order that a joint commitment be rescinded, the parties must concur in its rescission. If an individual fails to conform to a joint commitment, then, absent certain background understandings, she offends against all of the other parties to the joint commitment.

Amplification of this last point is important for a full understanding of the concepts of joint commitment and *collective belief*, that is, group belief on Gilbert's account of it.[12] The offense just referred to is a matter of violating an obligation to the other parties, who have rights to one's conformity to the joint commitment. This is a function of the joint commitment itself: a joint commitment obligates each party *to the others* to conform to it, at the same time endowing them with rights to such conformity.[13]

Once party to a joint commitment, a person is required to act in accordance with the commitment under penalty of rebuke from the other members of the group. Should someone indicate that he is about to fail to accord with the commitment, the other parties have the standing to demand that he conform after all.[14]

The notions of "rebuke" and "demand" here are strong ones. As just indicated, in order to rebuke someone or demand some action of him, in the relevant senses, one must have a certain standing or authority. Thus it is not the case that everyone who finds one's action distasteful is in a position to rebuke one for it, or that everyone who thinks one ought to perform some action can demand that action of one.

Further, there is a sharp distinction between having the standing to demand a certain action of another and being justified in making that demand. The same thing holds for rebukes. Sometimes one will have the standing to make a demand, but will be unjustified, all things considered, in making it. Perhaps the psychological make-up of the potential addressee makes that inadvisable, for instance. Or perhaps the otherwise acceptable action one has the standing to demand is not justified, all things considered, in the circumstances. In that case, presumably,

---

[12] Henceforth we use the phrase "collective belief" as short for "group belief on Gilbert's account of it."

[13] For discussion of the relationship of joint commitment and directed obligation, see Gilbert (2012) and elsewhere.

[14] Gilbert sees the following as "equivalents" in roughly the sense of the rights theorist Wesley Hohfeld: (1) A has a right against B to B's doing ϕ; (2) B is obligated to A to do ϕ; (3) A has the standing to rebuke B for not ϕ-ing; (4) A has the standing to demand that B ϕ (before the time for A's appropriately ϕ-ing has passed). See Gilbert (2012).

one will not be justified in demanding that action. Again, one might lack the standing to make a certain demand, though, if one had that standing, one would be fully justified in exercising it. Finally, one may be justified in pressuring someone to do something or in letting them know that one thinks poorly of their doing it, without having the standing to demand that they do it or the standing to rebuke them for doing it.

In the particular case of a collective belief, these consequences of joint commitment mean that if an individual is party to a collective belief and if she speaks in a way that contradicts the belief (without significantly qualifying her statement), then she has offended against the other members of the group.[15] In other terms, she has failed to fulfill obligations she has to them, and they have the standing to rebuke her for this failure. To see how this works in a particular case, imagine that the parents of a teenager disagree about when the teen's curfew should be. The mother believes it should be 9 p.m., whereas the father says 11 p.m. They compromise, and later tell the teen, "We think you should be home by 10." If the father then says, "Actually, 11 is fine," the mother would rightly feel affronted. In addition to any other reactions she may have when or after their son is present, she may well rebuke the father for speaking as he did, and he will understand that she has the standing to do so.

According to Gilbert, *any* group of people holding a collective belief counts as a social group. A group of people is a *social group* for Gilbert just in case the members are jointly committed in some way. In that case, she argues, each member appropriately thinks of the group as "we" or "us". She refers to social groups of this sort as "plural subjects". Two or more people constitute a plural subject of ϕ-ing, where ϕ-ing is the activity/belief/etc. about which they have a joint commitment.

Even though "social group" is a technical term here, it is intended to capture the same content as would the words interpreted informally, at least in a certain central sense.[16] Intuitively, the idea is that a collection of people becomes a social group when they openly decide to "join forces" in a coordinated action (say). Thus, suppose that Wolfgang comes across Alice struggling up a hill with her groceries. Wolfgang stops and offers to help; Alice accepts. Wolfgang takes a few of Alice's bags, and then the pair walks up the hill together. Now Alice and Wolfgang are jointly committed to espousing as a body the goal of carrying

---

[15] The kind of qualification at issue includes in particular one's using such a preliminary as "Personally, I . . .", that makes it clear that one is about to express one's personal belief as opposed to the collective's belief. See e.g. Gilbert (2000a).

[16] See Gilbert (1989), ch. 4, and elsewhere.

Alice's groceries up the hill. Suppose a few minutes pass and Alice sees her friend Jacques. Jacques stops and asks what is going on, to which Alice can justly reply, "*We* are carrying my groceries home." They are sharing the action in a way that they would not be if they both happened to be carrying groceries up the same hill at the same time. It is for this reason they count as a social group. Collective belief, or rather, the joint commitment to believe that p as a body, is just one example of the kind of joint commitment necessary and sufficient for social group formation.

As noted above, it is helpful to distinguish Gilbert's position from another possible view that is common in the philosophical literature. Gilbert's account of group belief is not "summative," in that Gilbert does not take statements of the form "G believes that X," where G is a group, to mean "all or most of the members of G believe that X," or any of the possible variations on that theme. Indeed, it is neither necessary nor sufficient on Gilbert's view for most (or even any—recall the curfew example) of the people in G to *personally* believe that X. A group can believe a proposition even if few or none of its members believe it, so long as the members are jointly committed in the right way.

It is of particular importance for what will follow that this joint commitment to believe that p as a body need not entail a commitment on behalf of each member personally to believe that p, or even personally to act as though he believes that p. Rather, the members of the group are to act as separate *mouthpieces* of the group, expressing the belief in that role in any setting in which they are acting in their capacity as group members. Conversely, if every member of a group happens to believe something, it does not follow that they believe it as a group, as it is possible that they have not yet jointly committed to believe it as a body.

As observed above, if Gilbert's analysis of sentences of the general type in question is correct, it should apply equally well to assertions of scientific consensus understood as ascriptions of belief to the relevant group. One can thus construe scientific consensus as a matter of collective belief. Scientific change, then, which amounts to moving from one consensus to another (at least according to a prominent vein in the history and philosophy of science), can be thought of as collective belief revision.

Before asking whether the features Smolin attributes to the string theory community are well explained by the suggestion that the community has one or more collective belief, it is worth pointing out some consequences that a consensus qua collective belief would have for a scientific community, on Gilbert's account. The most important of these all follow from the fact that the existence of a scientific consensus would imply that a scientific community is a social group with a joint commitment, which in turn implies that members of the community have obligations to behave in certain ways. In particular, this means that the members of a scientific community are obligated to act as mouthpieces of the group with respect to the consensus, or risk rebuke from other members of the community.

Expressing a contrary view—bucking the consensus—is an offense against the other members of the community, and threatens to put the contrarian outside the bounds of the social group.[17] So, irrespective of their personal beliefs, there are pressures on individual scientists to speak in certain ways. Moreover, insofar as individuals are psychologically disposed to avoid cognitive dissonance, the obligation to speak in certain ways can affect one's personal beliefs so as to bring them into line with the consensus, further suppressing dissent from within the group.

Finally, if scientific consensus is a set of collective beliefs, then effecting scientific change is not necessarily a matter of convincing a majority of individual scientists that a new view is correct. Instead, it is necessary to get the members of the community to jointly commit to believing, as a body, a new proposition. That is a particularly arduous undertaking, in that it requires one or more individual scientists at least temporarily to risk rebuke from their colleagues as they attempt to build public support for the new view, expressing support for that view in the face of a contrary consensus.

Considerations of cognitive dissonance and potentially conflicting commitments may have another consequence as well, regarding how scientists, both individually and collectively, deal with evidence for propositions that conflict with their consensus. Bringing up such evidence will itself have costs, in the form of possible professional rebuke, akin to those associated with outright denial of the consensus.

In some cases, these costs may prevent new results from being submitted by individual scientists or research teams to scientific peers for consideration, or from being selected by relevant individuals or committees for presentation to scientific peers.[18] In others, individual scientists or research teams may avoid pursuing potentially transformative research in the first place.

---

[17] A qualified expression such as "Personally, I have my doubts about the theory" may avoid outright default on an obligation, but it is likely to make one stand out as a potentially unreliable group member. See Gilbert (1987, 2000a).

[18] As may have happened in the case of the initial proposal of a bacterial theory of ulcers—now well entrenched. This was one of very few papers rejected for a gastroenterology conference when the received view was that ulcers were caused by other, including dietary, factors. See Gilbert (2000a), citing Thagard (1999), then in manuscript form, which to some extent inspired it.

Even when evidence against a consensus is found by an individual scientist, one might expect it to be ignored, suppressed, or explained away by its discoverer, since such evidence would force a psychologically unsustainable conflict between the scientist's commitment to act in a certain way and his (or her) beliefs concerning the epistemic warrant for those actions. In the case of evidence contrary to a consensus that is made public, the scientific community, also, may ignore, suppress, or explain it away. For instance, it may be collectively affirmed that a crucial experiment cannot have been properly done, or it may be assumed that facts that are not at odds with the consensus can explain it, however implausible such an assumption really is. For these reasons, scientific communities should be expected to hold certain beliefs in the face of considerable conflicting evidence, to the point of being, or at least appearing, irrationally dogmatic or epistemically irresponsible.

3. Kuhnian Paradigms and Gilbertian Collective Beliefs

Gilbert's account of scientific consensus and change can be brought to bear on what are sometimes called "two-process" views of scientific change, including the view developed and defended by Kuhn (1962).[19] On Kuhn's picture, scientific change (at least in "mature" sciences) occurs in two distinct modes.

One mode, which Kuhn calls "normal science," is characterized by the broad acceptance of a "paradigm." A paradigm, meanwhile, consists in (at least) a collection of theoretical commitments, acceptable research methods, and recognized problems of pressing interest. When working within a paradigm, scientists may apply their accepted methods to make incremental progress on the problems recognized by the community. This is change *within* a paradigm.

The other mode, meanwhile, which Kuhn calls "revolutionary science," consists in change *between* paradigms. That is, during revolutionary science, scientists reject a previously accepted paradigm and adopt a new collection of theoretical commitments, research methods, and important problems.

The connection to Gilbert's proposal can be seen by observing that the theoretical and methodological commitments associated with a paradigm may best be understood as a set of foundational collective beliefs of the community of

---

[19] The terminology of "one-process" and "two-process" views is due to Godfrey-Smith (2003). Aside from Kuhn, one might recognize Carnap (1956), Lakatos (1970), Laudan (1977), and Friedman (2001) as defending two-process views. Much of what is said here relating Kuhn and Gilbert could be extended to relate Gilbert's proposal with these other views.

scientists working within the paradigm.[20] In other terms, the collective beliefs in question create the overarching framework within which the work of this community is conducted. Revolutionary change, then, may best be conceived as a variety of collective belief revision, consisting in the rejection of a prior joint commitment with respect to one or more core propositions associated with a paradigm, and the institution of at least one new joint commitment that conflicts with these core propositions.

This way of thinking about Kuhn's views offers an explanation of one of the most striking and controversial features of normal science. Specifically, Kuhn argues that during periods of normal science, "anomalies," which are theoretical or experimental discoveries that are apparently incompatible with the central tenets of a paradigm, are either not recognized or ignored. In other words, during normal science, researchers seem to focus on evidence that appears to confirm the beliefs associated with the paradigm, and to disregard contrary evidence. Given that, on many views of evidence, high-quality contrary evidence should be taken to be of especially high value, this tendency to ignore contrary evidence may seem (at least) strange, and perhaps irrational.

Note, however, that if a paradigm is a collection of foundational collective beliefs of a scientific community, this attitude towards conflicting evidence should be expected. As discussed earlier, some degree of resistance to contrary evidence is predictable given the nature of the joint commitment constitutive of any collective belief whatsoever. That is so both for individual members or sub-communities of a given scientific community and for the community as a whole.

Given the foundational nature of the beliefs constitutive of a paradigm, one would expect an even greater tendency to resist contrary evidence than is present in every case of collective belief. Such resistance would be evident in, for example, the particular harshness of the rebukes meted out should one challenge these core collective beliefs. At the extreme, a given member of the community may be judged to have removed himself from the community—to have ex-communicated himself. After all, acceptance of a challenge to a given foundational belief is apt to bring down the whole edifice of beliefs within which this community has been working—perhaps for a very long time.

In sum, the apparent conservatism that Kuhn attributes to scientific communities may be explained by the general nature of scientific consensus understood

---

[20] In a more nuanced treatment one would most likely bring in phenomena other than group beliefs as Gilbert construes these, phenomena such as the joint acceptance of certain methodological rules or conventions, each of which can be construed in terms of joint commitment. (On the latter, see Gilbert 1989, ch. 6; 2013, ch. 9.) Doubtless, however, beliefs of one kind or another play a central role in any Kuhnian paradigm and, for present purposes, we write as if all of its elements are matters of belief.

as collective belief. This conservatism would be amplified given the centrality of the beliefs in question.

One might push this idea still further. Kuhn seems to suggest that, despite the apparent failures of rationality associated with ignoring or suppressing contrary evidence, the epistemological features of normal science provide a partial explanation of the *success* of science.[21] The idea, here, is that science is successful in part because of a distinctive kind of focused, collaborative research. And this sort of collaborative research is enabled by the existence of a paradigm insofar as a paradigm provides a stable shared agenda and collection of methods for realizing that agenda. The resistance of a scientific community to accepting changes in the paradigm thus provides a mechanism for preserving this collaboration.

Gilbert's account of group belief suggests a way to understand this mechanism. Indeed, it suggests a way of understanding any relatively long-term collaborative process.

Suppose one accepts that science, at least as understood in the modern period, is an essentially collaborative activity of relatively long duration, and that any such collaboration requires a framework of beliefs, concerning, at least, the nature and viability of its goals and the best way to achieve them. If these beliefs are conceived as group beliefs in Gilbert's sense, then they will be constituted by an appropriate set of joint commitments. As we have discussed, such joint commitments, once established, are apt to provoke resistance against anyone who is inclined to push for their rescission. Thus, the resistance to change on which Kuhn focused need not be conceived as a special feature of scientific communities. Instead, these may be seen as characteristic features of *any* long-term collaborative activity. The participants can be expected to resist change with respect to the framework of collective beliefs that help to define its goals and the means to be adopted to achieve them—its methods, if you will.

In the case of a long-standing scientific discipline, of course, there will be a special corpus of collective beliefs that represent not only the approved aims and methods of the enterprise, but an increasingly sophisticated body of theory, supported where possible with accredited empirical results, involving a host of linkages between foundational propositions and others. The collaborative

---

[21] Here we are largely setting aside the question of what it means to say that science is successful—that is, whether the success of science should be measured by the empirical adequacy of scientific theories, the truth of those theories, their explanatory power, etc. Kuhn's own notion of the success of science was deeply entwined with his notion of normal science as "puzzle solving." Roughly, Kuhn argued that science is very successful at solving the puzzles deemed to be important within a paradigm, and that the features of normal science described this sort of success. Our point here is merely that Kuhn gave reasons to believe that the conservatism of normal science may not be the impediment to the success of science that it would appear to be.

enterprise that is science in its various branches is, then, of particular interest to the collective epistemologist. The collective beliefs of a mature science go far beyond those that concern a set of goals and methods, or arise incidentally in the course of pursuit of a collective goal. They involve a multiplicity of highly articulated, closely interwoven, and mutually sustaining collective beliefs about the world.

Indeed, the point of the enterprise, prescinding from various epistemic cautions, is that those beliefs be true. Hence conservatism with respect to scientific paradigms, in particular, has its problematic side. Though it helps to provide a climate in which fledging theories can grow and flourish, it also helps to hide from view alternative theories that may in fact be better.[22]

4. The Rise of Strings

We now turn to string theory, and offer, first, a brief history of this theory. Today, string theory is a would-be "final theory," that is, a theory with foundational aspirations. It originated in the late 1960s, however, in a role rather different from its current one.[23] Initially, string theory was a phenomenological attempt to understand one of the four fundamental forces, known as the strong nuclear force.[24] The strong force acts on particles known as "hadrons," among which are the relatively familiar proton and neutron that form the nuclei of atoms. By the late 1960s, it was believed that these hadrons were composed of smaller particles, called "quarks," but their properties were not well understood. All that was known was that quarks must be, in some sense, "confined" to hadrons, since no one had ever observed any free quarks. String theory was supposed to explain this confinement via very small elastic bands (the strings) that bound the quarks together. Although the theory was fairly successful, its progress halted abruptly in

---

[22] Here there is a connection to recent work by Stanford (2006) on the problem of "unconceived alternatives." Stanford argues that scientists' systematic failure to identify alternative theories that can deal with available evidence as well as or better than current theory is the strongest threat to scientific realism. If the conservatism of normal science as we have described it here contributes to that failure to recognize alternative theories, then it bears severe epistemic costs. Of course, Kuhn was not a realist in the sense Stanford attacks.

[23] As mentioned in the introduction, the material here is derived from Galison(1995) and Smolin (2006); for additional perspectives, see Cappelli *et al.* (2012) and Dawid (2013). This section is intended as background, and not argument. If anything of the history presented here is contentious—aside, of course, from Smolin's claim that string theory has not accomplished its stated goals and should now be viewed as a failure—it is unintentionally so.

[24] As a matter of vocabulary, physicists use "phenomenological" to refer to models/theories intended to describe specific phenomena. The contrast class would be "fundamental" theories, which claim to be generally valid and universally applicable (at least in principle).

1973 with the remarkable experimental success of a competing theory of the strong force, known as quantum chromodynamics (QCD). As Galison points out, however, string theory had flaws of its own: in its capacity as model of the strong force, string theory predicted a new particle with no counterpart in the phenomena to be explained.

As it became clear that string theory was a dead end with regard to hadronic physics, committed devotees, convinced of the power of the theory's mathematical structure, looked for new applications. In 1974, John Schwarz and Joël Scherk, and independently T. Yoneya, proposed a reinterpretation of string theory. This new theory was essentially identical to the old one, except now the strings were 10^{20} times smaller. Instead of binding the constituents of hadrons together, strings were now proposed as the fundamental building-blocks of both elementary particles and spacetime itself. In this role, the unobserved particle that threatened to derail the hadronic theory could be interpreted as a carrier of the gravitational force. (The problem of finding an adequate quantum-mechanical theory of gravitation had proved remarkably stubborn, and any candidate for a particle corresponding to gravitation was considered a promising one.) In this new role, however, the theory had some worrisome and undesirable properties: for one, it predicted that the universe had at least six additional, unseen spatial dimensions. Whether for this reason or some other, over the next decade string theory was largely ignored by the physics community, aside from a small group of researchers.

All this changed in 1984, with such dramatic suddenness that the period is often referred to as the first superstring revolution.[25] The tipping point was a calculation by Schwarz and a young collaborator named Michael Green that appeared in August of that year. They showed that string theory lacked an inconsistency that had plagued other so-called unified theories then under consideration. The response was surprise, celebration, and a massive movement of physicists into the field. In 1984 there were about 150 articles published on string theory in total, about three times the average annual output in the previous decade. In 1985 the number was well over 400, and then over a thousand in 1986.

This explosive growth can be attributed to a variety of factors. One was that the then dominant theory—called the Standard Model, which included QCD as one of its two subparts—had been around for a decade and was doing too well. None of its predictions had been falsified and a good number had been confirmed to a high degree of accuracy. Wonderful as this sounds, it was widely believed that the

---

[25] "String theory" is now common shorthand for a theory that was known as "superstring theory" when it was first developed in its modern form, during the 1980s.

Standard Model could not be a fundamental theory, because it left too many of its own parameters unexplained; yet, without any experimental disagreement, there was little to point the way towards more fundamental theories.

After the Schwarz–Green calculation, string theory was a promising possibility in a landscape where all other options seemed exhausted. It was in this period, between 1984 and 1989, that string theory first rose to be the leading candidate for a theory of everything. Smolin also describes the mid- to late 1980s as the time when the string theory community began to exhibit the sociological features he highlights. (We will state and discuss these in detail below.) He writes of string theory: "It was the hottest game in town... Very quickly there developed a cultlike atmosphere. You were either a string theorist or you were not" (Smolin 2006, 116). One reason for this appearance of division, according to Smolin, was that string theory required new technical tools that most physicists would not have learned in graduate school. The investment of time and energy were risky, and were taken as evidence of one's commitment to the new project. Theorists who did not take the time to learn the new tools were viewed as either incapable of understanding the new developments (a stance common among younger physicists towards their elders, when the elders began to question the unconventional new research). And it was easy to tell who had devoted themselves to the new theory, because the research methods were distinct enough to distinguish "string theorists" from others, based solely on the sorts of papers they published.

It was also during this period that the first outspoken dissenters appeared. Among these critics numbered many of the most prominent theorists of the previous generations. Nobel laureate Richard Feynman, for instance, wrote in 1988, "[string theory] doesn't produce anything; it has to be excused most of the time. It doesn't look right." Another Nobelist, Sheldon Glashow, who was largely responsible for a big chunk of the Standard Model, wrote in the same year that string theorists "cannot demonstrate that the standard theory is a logical outcome of string theory. They cannot even be sure that their formalism includes a description of such things as protons and electrons. And they have not yet made even one teeny-tiny experimental prediction" (both quoted in Smolin 2006, 125). Howard Georgi, who with Glashow proposed the first Grand Unified Theory (and thus started the path of which string theory was supposed to be the end), wrote in 1989, "I feel about the present state of GUTs as I imagine Richard Nixon's parents might have felt had they been around during the final days of the Nixon administration" (quoted in Galison 1995, 392).

Interest in string theory calmed down somewhat during the early 1990s. Approximately 600 papers were published each year between 1990 and 1993—half the peak of 1,200 in 1987. Although many still found the theory to be promising, the wide interest of the late 1980s had revealed several new undesirable features. Most prominent was that there appeared to be a handful of *different* theories, all justly going by the name string theory, and between which there was no way to adjudicate. This was in conflict with one of the principal virtues many physicists hoped and expected string theory to have, namely, that its mathematical structure would lead to an essentially unique theory. Physicists felt that a highly constrained theory of this sort was desirable because, if confirmed, it would carry a sense of necessity with it. The world is the way it is, one might then say, because small changes (in the true theory) would lead to a mathematical inconsistency.

Soon physicists were able to classify five possible string theories. This number might have seemed manageable, except that to make any of these five theories physically acceptable, it was necessary to "compactify" (literally, roll up and hide from view) the six extra dimensions. By the late 1980s it had been observed that there were millions of consistent ways to do this, and picking the one that corresponded to the world appeared arbitrary and *ad hoc*. In other words, string theory appeared incapable of making substantive predictions at all, since the geometry of the theory was so radically underdetermined by known mathematical constraints that almost any possible experimental data was compatible with the theory. Twenty years later, these features continue to be among the ones that string theory's critics cite.

The field accelerated once again in 1995, however, when Edward Witten made a rather striking proposal. He said that although it seemed there were five different string theories, he believed it was possible that all of these were examples of a single underlying theory, with seven extra dimensions (instead of six). This new theory did not have strings in it, *per se*. Now it had two-dimensional surfaces. One of the dimensions of these surfaces was tightly wound in the new extra dimension, so that the objects would appear to be one-dimensional strings in a ten-dimensional space, much as in the older theory. Witten conjectured, and later proved, that the five string theories that were discovered in the late 1980s corresponded to different ways of winding the two-dimensional objects around in this eleventh dimension. He named the new theory M-theory. Once it was understood that the five theories were actually different parts of one theory, physicists' interests were reignited. (The additional problem of compactification, however, remained.)

This period, following Witten's proof and a handful of developments that followed quickly from it, is often called the second superstring revolution. Since then, string theory and its descendant, M-theory, remain hegemonic, despite the fact that some of the earliest concerns—about the prediction of unseen particles without any details of their properties; about the non-uniqueness of the scheme for hiding the theory's extra dimensions; and about the lack of connection with experiment—remain unresolved.[26] The mid-1990s saw a slew of popular books, by such authors as Michio Kaku and Brian Greene, that spread the word of the superstring revolutions to non-scientists, and helped create the sense that, despite its problems, string theory was already established science.[27] Now, however, many (perhaps most) outsiders to the theory, in both the public press and in other areas of physics, tend to be more critical. Despite the efforts of some string theorists, such as Leonard Susskind of Stanford and Lisa Randall of Harvard, to present new developments of the theory and sustain a feeling of hopefulness, an increasing number of physicists outside of the string theory community have come to believe that the grace period during which the theory's problems could be excused has ended. And yet it continues to be dominant within its sub-field of physics. In other words, from Smolin and others' perspectives as outside experts evaluating string theory, given the current state of the field, string theorists appear to be epistemically irrational in their continued confidence that string theory is the best available proposal for approaching questions in quantum gravity.[28] This is where Smolin suggests that sociology plays a role.

---

[26] This is not to say that nothing has changed in the last fifteen-to-twenty years. In particular, the interests of string theorists have shifted to new topics, including questions concerning the role of strings in cosmology and the early universe and the so-called AdS/CFT correspondence, which attempts to draw a connection between string theory and more traditional approaches to high-energy particle physics. Curiously, some string theorists have also had success in applying the methods characteristic of their discipline to problems in other, radically different areas of physics, including atomic and nuclear physics. (This latter work has led some string theorists to claim that string theory *has* made testable predictions. But they are not predictions concerning fundamental physics, the supposed domain of the theory. At best, predictions of this other sort provide evidence that some of the mathematical and physical reasoning used by string theorists is not inherently fallacious; it emphatically does *not* provide evidence that string theory is the correct fundamental theory of nature.) Importantly for present purposes, the *sociological* features of string theory that Smolin emphasizes were already well-established by the mid-1990s and, he claims, had not changed significantly by the time he wrote his book.

[27] Greene (1999, 213–14) in particular provides quotes from some of the same critics Smolin and Galison cite—including Georgi and Glashow—that are more conciliatory.

[28] String theory has not been without "outside" defenders—for instance, the particle physicist-turned-philosopher Richard Dawid (2013) has recently argued that not only is string theory not a failure, but its success should force a change in how we understand science.

5. Smolin's Sociological Critique: String Theory and Collective Belief

In his *The Trouble with Physics* (2006), Smolin describes seven features of the contemporary string theory community that are intended to explain why it continues to be the dominant candidate for a fundamental theory, despite the widely held view of experts outside the fold that the theory is no longer as promising as it once seemed.[29] We first state them in their entirety, and then we relate them to Gilbert's description of a scientific community holding one or more collective beliefs. Smolin writes that string theory has qualities of:

- *Tremendous self-confidence*, leading to a sense of entitlement and of belonging to an elite community of experts.
- *An unusually monolithic community*, with a strong sense of consensus, whether driven by the evidence or not, and an unusual uniformity of views on open questions. These views seem related to the existence of a hierarchical structure in which the ideas of a few leaders dictate the viewpoint, strategy, and direction of the field.
- In some cases, a *sense of identification with the group*, akin to identification with a religious faith or political platform.
- A strong sense of the *boundary between the group and other experts*.
- A *disregard for and disinterest in* the ideas, opinions, and work of experts who are not part of the group, and a preference for talking only with other members of the community.
- A tendency to *interpret evidence optimistically*, to believe exaggerated or incorrect statements of results, and to disregard the possibility that the theory might be wrong. This is coupled with a tendency to *believe results are true because they are "widely believed,"* even if one has not checked (or even seen) the proof oneself.
- A lack of appreciation for the extent to which a research program *ought to involve risk*. (Smolin 2006, 284, emphasis in original)

In what follows, we work with the hypothesis that string theorists are a group of people with collective beliefs concerning the fundamental nature of the world.

---

[29] We should emphasize that the claim is not that string theory has failed—insofar as it has—because of these sociological features. Instead, string theory has failed because it is not an adequate theory of nature. It continues to enjoy a privileged place in the physics community, however, for these sociological reasons; this continued dominance, meanwhile, has prevented other possible theories from receiving much attention, which, Smolin claims, explains why high-energy physics has stalled. Thanks to Gerald Cantu for pointing out this possible ambiguity.

(Given the connection we have noted with Kuhn's work, one might rephrase this as "a group of people working within a Kuhnian paradigm.") We will use the proposition "string theory is true" as shorthand for whatever those beliefs are supposed to be, setting aside the worry that "string theory" may not refer to a single, well-defined set of propositions. The actual views of string theorists are more subtle, and would involve a number of technical propositions and beliefs about the relation between these propositions and the world, plus opinions about the ontology of the world and the prospects for string theory's ultimate success. It is not important precisely what statements are part of the core set of collective beliefs of the community, so long as there *is* such a core set shared among all string theorists.

We will group the features that Smolin describes according to their explanation in terms of collective belief. We will argue that all of these features have natural explanations, given the assumptions made thus far concerning the string theory community and the nature and implications of collective belief. The central assumption regarding the string theory community for present purposes is its possession of a core set of collective beliefs—group beliefs in Gilbert's joint-commitment sense. Then, in the section "Seven *unusual* features?" we will turn to the question of why, if the features Smolin describes are precisely what one should expect of a scientific community, Smolin describes them as "unusual."

Features 2, 3, and 4: identification with the group and a boundary with other experts

Features 2, 3, and 4 of Smolin's description are direct consequences of collective belief. Holding a collective belief is a sufficient condition to constitute a social group, which means that string theorists can justly refer to themselves as "we" with regard to the consensus of the community. So it is unsurprising that the parties to the collective belief tend to have a sense of identification with the group holding the belief. After all, they are *members* of that group. The point will presumably pertain even more strongly to a population of scientists with many interconnected collective beliefs.[30]

Likewise, the sense of a boundary between the string theory community and individuals and other groups of a different persuasion follows from the string theory community's status as a social group. Insofar as the people who

---

[30] Can a given group have inconsistent beliefs? If so, it could indeed be described as a house divided. Supposing that this can happen within a given scientific community, it is clearly a special though interesting case that we shall set aside here.

collectively believe that string theory is true can refer to themselves collectively as "we," there is an available distinction between "us" and "them". And so features 3 and 4 can be expected of any social group.

Similarly, feature 2 should be expected of any social group holding one or more collective beliefs. A monolithic community with a strong sense of consensus is precisely what would characterize a group with multiple joint commitments to speak and act in ways expressive of particular beliefs—particularly if there is a large number of interconnected beliefs with a central core.

Indeed, Smolin's description of the early history of string theory suggests the process by which an initial joint commitment, out of which the social group arose, was formed. Early string theorists needed to devote significant amounts of time to learning new theoretical methods—different enough from other methods that Witten once described the theory as "a piece of 21st-century physics that fell by chance into the 20th century" (Cole 1987)—and it was clear who among the broader physics community had chosen to do so. As Smolin explains, physicists who *did* choose to learn string theory quickly came to view themselves in opposition to those physicists who did not.

Some physicists refer to those committed to string theory in terms appropriate to a plural subject: as Harvard string theorist Andrew Strominger put it later, reflecting on the early days, "We were once considered semi-crackpots working on some bizarre idea" (Johnson 1998). Their expenditure of time and energy, with consequences that could be easily observed by their peers, amounted to a public commitment on the part of those who made the investment. By the mid- to late 1980s, at least, it seems likely that through their interactions in the course of their work the relevant group of physicists had thus openly expressed their readiness *jointly* to commit to work as a body on string theory, in conditions of common knowledge, so that the conditions for a Gilbertian social group obtained. As described above, it was during the same period that a consensus concerning the basic propositions of string theory was first arrived at by those physicists jointly committed to working on string theory, and that Smolin's feature 4 became apparent.[31]

---

[31] As indicated in n. 2, the distinction (or rather distinctions) between belief and some form of "acceptance" made by various philosophers is not strictly germane here. *Mutatis mutandis*, the points made here about collective belief can be made about collective acceptance, on whatever construal. The question whether it is better to characterize string theorists as collectively believing as opposed to collectively accepting the basic propositions of string theory is an interesting question that we shall not pursue. For some, it will hang on whether they are collectively agnostic as to the truth of these propositions, as opposed to the desirability of supposing their truth for the purposes of scientific inquiry.

Features 5, 6, and 7: disinterest in other ideas, tendency to interpret evidence optimistically, and poor appreciation of risk

Features 5, 6, and 7 follow only slightly less obviously from Gilbert's account. There are two things to say here: the first is to explain why these features might *appear* to be true of the string theory community to an outside expert like Smolin, and the second is to explain why they might in fact come to be true of the members of the community.

As parties to a joint commitment, members of the string theory community are obligated to act as (for short) mouthpieces of their collective belief. This means that if they speak in a way that contradicts the consensus, without significant qualification, they risk offending against other parties to the collective belief. As it is difficult (and professionally unwise) to offend against other members of one's community, one expects parties to any collective belief to refrain from speaking approvingly about ideas that contradict or challenge the consensus.[32] Note that, no individual can rescind the commitment alone, even in the face of overwhelming evidence against it—at most, she can choose to violate the commitment and risk rebuke. The collective belief entails an associated obligation to express skepticism about evidence or viable alternative views that might instill doubts about the consensus. Likewise, any evidence that can be taken in favor of the collective belief is to be treated with approval. Someone who is obligated to speak as (for short) the mouthpiece of a group is naturally inclined to emphasize evidence in its favor—that is, to take an appropriate rhetorical stance in evaluating and communicating the evidence. So features 5 and 6 seem to be predictable on Gilbert's account. Or at least, it is to be expected that an outsider would characterize a group with a collective belief as having features 5 and 6.

As Gilbert argues (2000*a*) and as we briefly noted above, features such as these may run deeper still. When a person is obligated to speak or act in certain ways, it can affect his private thoughts, inhibiting him from pursuing spontaneous doubts about the group view or from fully accounting for available evidence. Even though being party to a joint commitment to believe that *p* as a body does not require one to in fact believe that *p* individually, it is awkward and often difficult to believe one thing and say another with conviction. Whether consciously or not, this difficulty can impede a party to a collective belief from exploring possibilities that seem likely to lead to a contradiction between one's personal beliefs and the

---

[32] This is most likely true even if the approval is qualified as in "Personally, I approve...." See e.g. Gilbert (2000*a*).

belief referenced in the pertinent joint commitment. An individual's personal beliefs, then, are liable to be strongly influenced by considerations directly arising from the collective belief. And so the collective belief can in fact change how an individual will react to and interpret new evidence in such a way that features 5 and 6 become true of the members of a group with a collective belief.[33]

To explain how the appreciation of risk in a research program relates to these considerations, and to the obligations arising from joint commitments in particular, it is important to say just what Smolin means by risk. Risk-taking scientists "invent their own directions" and "tend to provoke strongly polarized reactions" (Smolin 2006, 342). Risk-taking within a research program amounts to exploring ideas that oppose the "entrenched approaches" on one's own initiative (p. 294). That is, risk, according to Smolin, amounts to systematically and sustainedly bucking the consensus—precisely what one cannot do as a party to a collective belief.

On this account of riskiness in research, an obligation to endorse string theory makes risk-taking even more unlikely, since if one succeeds in taking risks, one violates the applicable joint commitment and may well be excluded from the social group by one's fellows, who take one to have indicated one's own readiness to be so excluded. And so it is unsurprising that Smolin finds few risk-takers within the string theory community. One cannot take risks as Smolin understands them and yet still be sure of retaining one's membership of the social group.

Smolin's biography is telling here. He worked successfully as a string theorist for many years, before deciding to explore new possibilities; now, by his own lights, he is no longer a member of the string theory community.

Feature 1: tremendous self-confidence and sense of eliteness

Tremendous self-confidence or the sense that the group holding a belief is somehow elite are not features that Gilbert has previously discussed in relation to her account of collective belief. But there are two considerations that are in the spirit of Gilbert's account that lead one to expect string theorists to exhibit this feature.

The first consideration is an explanation of the appearance to an outside expert of tremendous (or irrational) confidence in the beliefs collectively held. Take Smolin as an example. He understands string theory, its consequences, and the evidence for and against it as well as any string theorist. On the basis of this

---

[33] This phenomenon of avoiding cognitive dissonance is not pure speculation. For instance, see Festinger and Carlsmith (1959), and the subsequent literature on forced compliance.

knowledge, he has determined that the status of string theory is at best uncertain. And yet members of this community speak as though string theory is true. We may suppose that, for the reasons described above, they inure themselves against evidence and ideas that might conflict with their collective belief. An outside expert, however, might reasonably assume that they are accounting for all of the evidence he is (especially evidence he might present in argument). In other words, the apparent irrationality of the string theory community—its unwillingness to update its beliefs in an epistemically responsible way, according to some outside experts—is naturally interpreted by outsiders as certainty bordering on dogmatism—or even as hubris.

Once again, as with features 5 and 6, this effect may also run deeper.[34] When a member of a social group with a collective belief speaks as a mouthpiece of the group, she is acting on an authority partially independent of and likely more significant than her own. She speaks for the group, and the views that she espouses have the blessing of the group's other members. If we add to this that there are distinguished intellectuals and academics for whom the speaker has great esteem among the members of the group, that would seem to make the effect only more prominent. When a string theorist says that string theory is true—or unassailable—she can do so with confidence derived from the understanding that she speaks with and for these distinguished members of the group. Moreover, members of the group see other string theorists behaving similarly. When they express the views of the group amongst themselves, they do so assuredly and with the gravitas that comes from believing they will not be contradicted. Seeing this confidence can be impressive: the members of the group appear to have authority and expertise. Each member of the group, seeing the other members behave thus, can easily come to believe that the group is populated with elite experts on the subjects that the collective belief concerns.

Seven unusual *features?*

As we have just seen, all seven of Smolin's features appear consistent with Gilbert's account. But there is still one aspect of Smolin's description of these features that does not appear to mesh. Specifically, if, as Smolin says, string theory is *unusual* in having these sociological features, then Smolin's view is in tension with the idea that joint commitments of the kind that constitute collective beliefs are likely to be present in *all* scientific communities. So why does Smolin claim that the features he describes are unusual, if in fact they obtain for any group holding a collective belief?

---

[34] Thank you to Cailin O'Connor for pointing out this consequence of Gilbert's view.

One way to resolve the tension is simply to suggest that what makes string theory unusual in this regard is a matter of degree—that is, perhaps the manifestations of the string theory community's joint commitment are particularly emphatic, for reasons independent of their group belief. On this suggestion, scientific communities may often have just the sociological features Smolin describes, and may even be monolithic in general. But for various reasons, string theory has turned out to be an especially striking case. (Indeed, Smolin describes string theory as "unusually monolithic, with a strong sense of consensus." One might take this as an acknowledgement that *all* scientific communities are monolithic to some degree, and all communities have some sense of consensus; string theory is merely an extreme example.)

A second possibility—compatible with the first—is that Smolin considers these features striking (or rather, they *appear* unusual to him) not because they are in fact unusual, but rather because Smolin and his collaborators stand in a different relationship to string theory than they do to other areas of physics. One might even think that Smolin's position with regard to string theory is, historically speaking, an unusual one for *any* prominent scientist to occupy.

The idea is that it is uncommon for the physics community to fracture into such clearly defined sub-groups with conflicting collective beliefs. More generally, it is unusual for a scientist to be a leader in his or her field—a fully competent expert—who nonetheless stands outside of a given collective belief within the field. But if and when this does happen, experts in the disagreeing communities would have a perspective that would make the features of the other communities—features such as those Smolin describes—more transparent, since such experts are well equipped to evaluate the evidence for and against a particular theory without being party to the consensus. If this is right, then string theory would seem quite different to Smolin than other areas of physics.

One can connect this point with the discussion of Kuhnian paradigms above. There we suggested that paradigms might be thought of as collections of fundamental collective beliefs, that is, beliefs to which the members of a given community of scientists are jointly committed. The apparently irrational features of normal science, then, are just consequences of the existence of core joint commitments obligating scientists to speak and act in certain ways on pain of severe rebuke or even excommunication. In these terms, Smolin's perspective on string theory is that of an expert in an area of science that is divided into groups with different paradigms, who stands outside of one of these paradigms, peering in.[35]

---

[35] Kuhn believed there could be at most one paradigm in a field at a time—a view famously challenged by Lakatos (1970) and Laudan (1977). So there is something un-Kuhnian about the

From this perspective—not so different, perhaps, from Kuhn's own perspective in *The Structure of Scientific Revolutions*—certain epistemic pathologies that most scientists never notice are cast into stark relief.

Of course, on Kuhn's view, there is nothing unusual about science within a paradigm. It is *normal science*, after all. But the fact that normal science has certain features does not mean that those features are obvious to working scientists. Quite the contrary. If anything, what is unusual is the ability to recognize these features at all, since to do so, one needs to be able to step outside of one's own particular corner of normal science and its associated paradigms and examine it from this external perspective. We suggest that this, in effect, is what Smolin has done. He has stepped outside the string theory community and is able to view it from an uncommon perspective, that of an informed outsider.

6. Conclusion

We have argued here that, given (1) Gilbert's joint commitment account of group belief, and (2) the hypothesis that the consensus within the string theory community can be properly construed as an example of group belief in Gilbert's sense, the seven sociological features that Smolin attributes to string theory are precisely what one should expect of a group holding relevant group beliefs. In particular, we have observed that, apparently contra Smolin, and consistent with a Kuhnian approach to normal science, these features may not be *unusual* after all. What *may* be unusual, however, is that such a circumscribed social group has come to exist as a sub-group within the physics community. Particularly, it may be uncommon for there to be practitioners in a given field with full accreditation and expertise who are not party to the joint commitments of the relevant scientific community. Then it would only be because Smolin occupies this "outside expert" vantage-point that the features of string theory appear unusual to him.

These considerations suggest that, at least in some cases, the social features of scientific communities are well explained by supposing that scientific consensus, both with respect to paradigms and otherwise, can be understood as a variety of group belief in Gilbert's sense, with its attendant obligations.

---

suggestion that there are multiple competing paradigms in quantum gravity. Thus, one might take the fact that Smolin does seem to stand outside of string theory when he criticizes it as evidence that Kuhn was mistaken, or perhaps as evidence that quantum gravity is in a period of crisis, during which paradigms are permitted to fracture, on Kuhn's account. One way or the other, however, it is fully consistent with Kuhn's views that Smolin's perspective is *unusual*, insofar as he has the expertise to evaluate string theory without being committed to the paradigm.

Before concluding, we would like to emphasize the following. As a matter of logic, that a given scientific community believes, in Gilbert's sense, that p, implies neither that p is true nor that p is false. Again, it does not imply that the community's belief that p is well grounded empirically, or that it is not. The most that can be said along these lines is that *if* the core beliefs of a scientific community are false, or poorly grounded, there are the noted barriers to that community's coming to reject these beliefs, *in spite of* their falsity or poor grounding. That is all that this chapter means to imply with respect to the string theory community, or any other.

11

Collaborative Research, Scientific Communities, and the Social Diffusion of Trustworthiness

Torsten Wilholt

1. Introduction

When we invest epistemic trust in results of scientific research, our trust is often directed at a collective body rather than at a single researcher. As an empirical observation, this should be quite uncontroversial. In this chapter I will argue that this empirical fact is not just due to convenience or habit: with regard to scientific information, every reasonable assessment of trustworthiness *must* attend to collective bodies rather than only to single researchers.[1]

Claims very close to this one have been defended with good reasons before, in particular by Deborah Tollefsen (2007),[2] but I will attempt a new approach to the problem. The novelty lies in the precise nature of the arguments that I will use to support the claim and, perhaps more importantly, the kinds of collective bodies that I have in mind.

In a first step I will appeal to the collaborative nature of most of today's scientific research. The trustworthiness of a collaborative research group does not reduce to the trustworthiness of its individual members. Again, this is very much in line with observations offered by other social epistemologists. But I will

---

[1] Trustworthiness is commonly attributed to either a piece of information or a source of information. I take it that the latter is the more basic kind of trustworthiness attribution, because usually the attribution of trustworthiness to a piece of information at least partially derives from an attribution of trustworthiness to its source. In any case, in this chapter, when I use the word trustworthiness I mean the trustworthiness of a source.

[2] While she has framed her arguments in terms of the question of what constitutes the testimony of a group, the above claim about trustworthiness could be regarded as a corollary of her thesis.

try to point out a few aspects of the nature of collaborative research that I think deserve special attention in this context.

Furthermore, I am going to argue that the social diffusion[3] of trustworthiness goes even further and extends beyond the level of collaborative groups: our epistemic trust is, and needs to be, also directed at entire research communities defined by shared methodological standards. As long as we regard methodological standards only as means of codifying and putting on record the procedures that are most suitable for arriving at reliable results, this sort of trust in research communities might appear to be merely a practical contingency. After all, collaborative groups or even individual researchers might be regarded as "in principle" individually responsible for finding out by themselves which methods are the most reliable ones. However, I am going to argue that important methodological standards are in many cases solutions to problems of coordination rather than optimization. They are thus conventional and irreducibly social in character.

I am also going to argue that the conventional standards by which research communities are defined are crucially important for those features of the research that determine its trustworthiness. Therefore, any reasonable assessment of trustworthiness must involve a focus on the research community.

2. The Collaborative Nature of Scientific Research

Collaborative research is on the rise. An indicator of this is the increasing number of authors per scientific paper. Co-authorship is a symptom of the collaborative nature of research that is particularly important in our context, because a published paper is arguably still the most important medium through which a collaborative research group functions as a provider of information (und thus becomes the potential recipient of epistemic trust). Extreme illustrations of the phenomenon of co-authorship are easy to find. An article on one of the particle detectors of the Large Hadron Collider lists 2,926 authors (The ATLAS Collaboration, G. Aad, E. Abat, *et al.* 2008). A paper on a large clinical trial comparing different thrombolytic drugs for the treatment of heart attacks cites the name of the collaborative research group as author; 972 investigators are named in an appendix to the paper (The GUSTO Investigators 1993). Multiple authorship is a general trend in many disciplines: In the United States the average number of authors per paper in the medical sciences increased from 3.7

---

[3] The phrase "social diffusion" and its use to indicate non-supervenience on the properties of individuals is shamelessly lifted from Goldberg (2006).

to 6.0 in the twenty years leading up to 2010; in physics the number more than doubled in the same time (from 4.5 to 10.1), in astronomy it more than quadrupled (from 3.1 to 13.8).[4]

When scientific results are generated through collaborative research, the trustworthiness of the group cannot be reduced to the trustworthiness of its members. To see this, first note that the trustworthiness of a group is certainly not simply a function of the average trustworthiness of its members, or of the trustworthiness of its least or most trustworthy member. Familiar arguments from social epistemology (Tollefsen 2007; cf. Surowiecki 2005) make this clear: The competence of a group can exceed the competence of even its most competent members, as can be illustrated by Condorcet's jury theorem and other considerations pertaining to the wisdom of crowds. Similarly, no single member of a collaborative research group has all the information that constitutes the basis for the competence of the group as a whole.[5]

Taken by themselves, these points might seem to leave open the possibility that the trustworthiness of a group may as a matter of fact somehow supervene on the trustworthiness of its members (if not by way of any simple and straightforward functional dependence). This might in fact be the case in a group that somehow forms its group testimony by a method of aggregation from the views of its members. But in many interesting cases, and typically in the case of a scientific research group, that is not so. In our context, an additional reason why the trustworthiness of the group does not reduce to the trustworthiness of its members should receive particular emphasis: A group's trustworthiness crucially depends on its social organization. More specifically, this concerns facts about how information is distributed among its members and processed within the group (including practices of deliberation among its members) and how epistemic trust *within* the group is enabled and maintained.[6]

---

[4] Numbers from National Science Board 2012, 5–36, fig. 5-24; see the underlying data at <http://www.nsf.gov/statistics/seind12/c5/fig05-24.xls> (accessed 25/3/2015).

[5] The last point is particularly important for groups that bring together expertise from different areas, sometimes different disciplines. The combined competence of a research group is often much greater than the competence of an individual scientist could ever be, as philosophers of science have often observed (Hardwig 1985; Thagard 1997; Wray 2002, 2006).

[6] Among these features of a group I would also count the degree to which the individual members of the group are approaching the group's epistemic objectives from slightly different angles. As philosophers of science have emphasized (esp. Longino 1990, 2002), the mechanism of mutual criticism is highly relevant for scientific knowledge generation. The diversity (or, as the case may be, homogeneity) of a group with regard to background assumptions, methodological preferences, and the like is thus likely to play a role for its overall trustworthiness as a provider of robust scientific results.

These observations have important consequences for the practical problem of assessing the trustworthiness of a group. With regard to the possibility of establishing and maintaining trust in the kind of groups that we encounter in the sciences, two features of present-day collaborative research deserve particular attention.[7] First of all, collaborative experiments or studies are often multi-site, that is, they are spatially dispersed and involve many institutions. This limits our ability to assess the trustworthiness of the group with the aid of institutional indicators (such as reputation and track record of institutions). The spatial dispersion, in connection with the size of the group, may also present difficulties for discerning and understanding the social organization of the group. The social organization of a large randomized clinical trial, for example, may often not be transparent even to most of the participating researchers themselves. Secondly, many groups exist only for the duration of a single experiment or study (if by "group" we mean the collective body that produces a given piece of scientific information). This makes it impossible to assess the trustworthiness of such groups with the aid of their track record.

It would seem that both these points refer us back to an assessment of the trustworthiness of a group's individual members. After all, with individual scientists we *can* use cues like their institutional affiliations and other indicators of their standing within the community as indirect evidence for their trustworthiness, and we can try to assess their track record. But, of course, the arguments we have considered just before remain unrefuted: The trustworthiness of a group cannot be derived from the trustworthiness of its members.

In its present form, the collaborative character of scientific research therefore presents us with considerable difficulties when it comes to assessing the trustworthiness of scientific information. Estimating the trustworthiness of a collaborative group as a source of information can be very taxing even for scientific peers, and will usually overtax the abilities and capacities of extra-scientific users of scientific information.

3. The Level of Research Communities

The previous section has revealed problems for a case-by-case assessment of the trustworthiness of particular sources of scientific information. This already seems

---

[7] Both these features and their epistemic significance have been stressed by Rebecca Kukla and Bryce Huebner in their paper "Making an Author: Epistemic Accountability and Distributed Responsibility," presented in 2011 at both the SPSP conference in Exeter and at the EPSA in Athens; cf. Kukla (2012); and Winsberg, Huebner, and Kukla (2014).

to provide a good pragmatic explanation for the phenomenon that epistemic trust in scientific results is often grounded in an attitude of trust towards science as a whole, or towards particular research communities.

However, as noted at the outset, I want to argue for a stronger claim than can be supported by such pragmatic considerations: A reasonable assessment of the trustworthiness of a source of scientific information *must* draw on an assessment of the trustworthiness of the pertaining research community as a whole. My conviction about this stronger claim rests on additional reasons that are by and large independent from the considerations pertaining to the collaborative nature of scientific research, as will become clear in what follows.

Research communities in the sense of this chapter are constituted by shared methodological standards. By methodological standards, I mean a wide range of guidelines and instructions for scientific practice that come in differing degrees of explicitness, generality and binding force and can govern all kinds of steps in the research process, from experimental design to data analysis to the dissemination of information. The intended scope of my use of the concept can best be illustrated by way of example. One methodological standard that is widely shared by several research communities is to regard 0.05 as the highest reasonable significance level in significance tests. Another standard, far more restricted in the scope of its application, is the following: in animal experiments in toxicology, you should always have a positive control group in order to test whether your experimental setup is sensitive to the kind of intervention you are investigating (cf. NTP 2001, vii). The biomedical research community has by and large adopted the following rule, which should be regarded as a standard concerning the dissemination of research information: before you start a clinical trial, you should register it with a public repository, including information on endpoints and study design (cf. ICMJE 2014, 11-2, WMA 2008, 3). As a fourth and final example, consider the following *proposed* standard, which was recommended in a discussion paper resulting from a workshop held by the German Association for Plant Biotechnology: in biosafety research on possible interactions between genetically modified organisms and their environment, every experiment should start from a scientifically based causal hypothesis.[8] The examples illustrate how standards range from the highly specific to the near-universal, as well as from the uncontested common ground to the controversial suggestion.

---

[8] The discussion paper "Biologische Sicherheitsforschung an gentechnisch veränderten Pflanzen," written by Inge Broer and Joachim Schiemann in October 2009, was circulated widely in the plant biotechnology community and publicized via the website of the German federal ministry of education and research; the proposed standards continue to be discussed, cf. Broer (2012).

Obviously, research communities shape the trustworthiness of scientific information by means of the standards they implement. In that sense, our trust in a piece of scientific research may at least be in part directed at the respective research community and their collective competence in setting the appropriate standards. But so far as this consideration goes, this might still be considered to be just a contingent matter. It might seem that that conclusion could be arrived at in the following way: in principle, there exists *the* most trustworthy scientific method for any given problem. The trustworthiness of any particular piece of information depends on whether it was arrived at using the most trustworthy methods. The role of research communities and their institutional organs (like working groups, peer review panels, and the like) is only to formulate guidelines that help individual researchers to identify the most trustworthy procedures. The collective efforts of research communities thus help to avoid the unnecessary duplication of individual cognitive efforts for identifying the most trustworthy methods, but their contribution is in no way essential or indispensable. Again, the community level would bear its trust-enabling role only for pragmatic reasons.

I think that the first step of this line of reasoning already contains a mistake: typically, there is no such thing as the objectively most trustworthy method for a given problem. I therefore hold that, in many important cases, methodological standards do not simply codify a method that can be regarded as the objectively most trustworthy one. Instead, they are conventional in character and thus irreducibly social.

My take on standards thus runs counter to a widespread view on methodology and epistemic quality management in general. According to this view, we can pass a purely epistemic judgment on methods or types of inquiry by telling how good they are at getting at the truth, and this decisive quality of a method is measured by its reliability. The most trustworthy method for tackling a given problem would therefore appear to be the most reliable one. In the following section I will argue that reliability cannot play the role it is assigned by this view. Methodological choices are underdetermined by the aim of arriving at reliable results.

4. Is There Such a Thing as the Reliability of a Method?

My argument starts from a very down-to-earth reflection that is familiar to every methodologist. In every empirical investigation there are at least two different kinds of way in which the investigation can go wrong: it can lead to a result which indicates that p, while p does in fact not obtain (call this a false positive); or it can fail to lead to such a result, while p is in fact true (false negative). If you are conducting an open investigation into the question of whether p, you do not yet know whether the true result is positive or negative. But the reliability of positive results and the reliability of negative results of one and the same method of inquiry can differ widely. For example, if a rapid HIV test was performed on a *random group* of German citizens, negative results would have a reliability of almost 100 per cent, while positive results would have a reliability of well under 50 per cent (that is, almost 100 per cent of those who would receive a negative result would in fact be free of the HIV retrovirus, while the proportion of HIV carriers among those who would receive a positive result would be under 50 per cent). So if you wanted to judge a method by its reliability, you would either have to pick one of the two types of reliability, or you would have to use an aggregate measure of its reliability. Let us consider these possibilities in turn.

Taking just one of the two measures by itself—for example, the reliability of positive results—and regarding it as indicative of the epistemic quality of a given method of inquiry is highly problematic, and not only because it seems arbitrary as long as you do not yet know what the result is going to be. A type of inquiry can only be relevant with regard to the question whether p if it gives a signal in support of p under certain circumstances and does *not* do so under certain other circumstances. Taken by itself, a high reliability of positive results is easy, almost trivial, to come by—for example, by designing a method that almost never produces positive results (only in obvious cases). For instance, the method of only identifying a tree as an oak tree when it is actually now carrying acorns has a high reliability of positive results, but it is not an overall commendable epistemic strategy in most situations. The same holds for the reliability of negative results if taken by itself. The method of always and only identifying a given tree as an oak tree if it is not carrying fir cones yields highly reliable negative results, but it is obviously epistemically deficient.

One might be tempted to conclude that what we obviously want is some kind of combined reliability. Let the epistemic quality be measured by the reliability of all results, no matter whether positive or negative. I will call this measure—that is, the proportion of truth among the positive and negative results taken together—a method's "overall reliability." However, overall reliability in this sense can be very misleading about the epistemic quality of a method. In many contexts, it is the sensitivity to a particular kind of case that is relevant to the epistemic quality of a method of inquiry, and not its reliability across all kinds of situations. This point is best demonstrated by considering a concrete example.

Consider, again, the example of the rapid HIV test, this time performed on the general population of Germany. The numbers I will quote in what follows are simplified but realistic estimates of the likely results of such an application,[9] based on the known sensitivity and specificity of such tests and current estimates of the prevalence of HIV in the German population, according to which an estimated 40,000 people have the virus. Of these, virtually all would be expected to test positive due to the very high sensitivity of the test. However, due to the very large number of healthy people tested (and due to the fact that the test's specificity is almost, but not quite, as good as its sensitivity), there would probably be around 239,880 cases of false positive results, although an overwhelming number of 79,720,120 would correctly receive a negative result. With 239,880 false results out of 80 million, the overall reliability of the test calculates as 0.9970.

Now compare this to the following fictitious placebo test, which simply always returns a negative test result, come what may. Applied to the German population, this would result in 40,000 false negative results (viz., all the results of the people who are actually carrying the virus), with the remaining 79,960,000 correctly receiving a negative result. With only 40,000 false results, this "test" would obviously have a higher overall reliability, namely 0.9995, than the actual rapid HIV test. Nonetheless, it is intuitively clear that the placebo "test" is epistemically completely useless, whereas the rapid HIV test is not.

We are therefore confronted with an example in which a method with lower overall reliability is epistemically preferable to an alternative with higher overall reliability. The insight can be generalized by identifying kinds of reasons that can under certain circumstances justify a preference for a method with lower overall reliability. Such a reason exists, for example, when you aim to avoid both types of mistakes (false negatives and false positives), but one type is particularly important to avoid. (Identifying individual cases of an infectious disease is a case in point, if the aim of limiting the spread of the disease is of particular importance.)

Another example of such a reason is given when you are searching for a phenomenon of very low prevalence. Think of a researcher collecting specimens in the field. If the target subspecies she is investigating is very rare, she has to take particular care in avoiding false negatives when she picks up each individual and decides whether or not it belongs to the subspecies in question and should

---

[9] Here, as in the earlier passage about the rapid HIV test, I am calculating for a hypothetical test with a sensitivity of 100% and a specificity of 99.7%. A recent WHO report gives these numbers as results of a laboratory assessment of one commercially available rapid diagnostic test (WHO 2015, 23). Other such tests were assessed with very similar results (with slightly worse figures for specificity in some cases, slightly better ones in others) (WHO 2015, 23–5). My hypothetical test by far exceeds the minimum requirements for HIV tests set by the WHO/UNAIDS, which are 99% sensitivity and 98% specificity.

therefore be collected. (On closer reflection, this second type of reason is thus a special case of the former.) The "method" of just never putting anything into the jar might very well have higher overall reliability than her careful scrutiny, but it is also obviously unsuited to get any investigation started. The example helps to point out that there may be a variety of reasons why you want to avoid one particular type of mistake. In particular, the possible extra-scientific consequences are not the only potential source of such a concern. Even if you are abstracting from such consequences, as the specimen-collecting researcher might very well be, differences in seriousness between false positive and false negative mistakes can be relevant and characteristic of the kind of inquiry that you are engaged with.

A preliminary conclusion from our reflections on reliability so far can be formulated as follows: a meaningful assessment of the epistemic quality of a given method must at the very least account for *both* the reliability of positive results *and* the reliability of negative results (at the same time, but separately). Note that these two types of reliability are in systematic tension: you can typically increase one at the cost of the other without increasing the data input or otherwise investing more effort. Therefore, every methodological decision involves a trade-off between them.

Can the two dimensions we have identified so far (the reliability of positive results and the reliability of negative results) *in combination* be considered sufficient for characterizing the epistemic quality of a method or type of inquiry? To see that they cannot, it is now important to move away from examples of standardized testing procedures.

Let us now consider types of inquiry for which one can draw a line between cases where the method produces a *negative result* and cases where there is *no result at all*. Many standardized procedures always have a result that is regarded as either positive or negative. But this is not so for many cases of scientific inquiry that do not just consist of applying a standard procedure. They often end in an inconclusive result, or lead to the resolve that more work is needed before a result of the inquiry can be declared. In practice, the line between negative results and no-result cases may be fuzzy in some cases, but for many actual types of scientific inquiry established and meaningful uses of this distinction exist. Methods or types of inquiry for which this holds differ not only with regard to how reliable their positive results are and how reliable their negative results are, but also with regard to how likely they are to produce any result at all, given a certain amount of effort expended.

This enables us to identify an important additional dimension that is relevant to how good a method is at getting at the truth, which I shall call the investigation's power. Power in this sense is the rate at which a method or type of inquiry generates definitive results, given a certain amount of effort and resources.[10]

A little reflection makes clear that even the consideration of both the reliability of positive results and the reliability of negative results cannot result in a significant measure of epistemic quality unless power is taken into account as well. In the absence of constraints on power, high reliability of both types of results usually comes cheap. For example, if we lowered the figure for the highest reasonable significance level that justifies the label "statistically significant result" from 0.05 to 0.01, we would have much more reliable results. But we would also have dramatically fewer results. The aim of getting at the truth implies not only that we want reliable results, but also that we want results.

The resources that we are able to dedicate to inquiry are always limited, which is why a method's power crucially matters to our epistemic aims. Similar to the two types of reliability, methodological choices also involve a trade-off between *both* types of reliability on the one hand, and the power of the investigation on the other. As an illustration, compare reliance on purely observational data in medicine with exclusive reliance on large randomized controlled trials (RCTs). You may want to ask: which of the two strategies is more properly geared towards the truth? But this question as such cannot be answered, because both trade off power and reliability in different ways. RCTs (let us grant) are more reliable both in positive and negative results, but much less powerful: observational data give us more results for the same amount of effort and resources. So the question which one of them is more properly geared towards the truth makes no sense, because both reliability and power are important dimensions of what it means for a method to be geared towards the truth. As an aside, one of the problems with the hierarchy of evidence as it is advocated by the movement for "Evidence Based Medicine" (EBM) seems to be that it gives all the attention and priority to reliability and none at all to power.[11]

What is desirable in a method of inquiry (from an epistemic perspective) can only be captured by considering all three dimensions: the reliability of positive results, the reliability of negative results, and the method's power. For each method, these three magnitudes form a triple that I will call the inquiry's *distribution of inductive risks*, or "DIR" for short.

---

[10] "Power" also is the name that Alvin Goldman (1992, 195) uses to identify a similar characteristic of a method, namely the rate at which it produces *true* results. Goldman's category of power thus already combines what I propose to call power with reliability.

[11] For critical methodological views on EBM, see Worrall (2002); Goldenberg (2006); Borgerson (2009).

DIR = ⟨reliability of positive results, reliability of negative results, power⟩.

This label takes up the talk of different types of inductive risk that was first introduced by Carl Hempel (1965, 91–2) and recently revived by Heather Douglas (2000, 2009). It extends the concept of inductive risk to include not only the risk of false negative mistakes and the risk of false positive mistakes but also the risk of ending up without any result at all.

If DIR is needed to provide a description of those features of a method that are relevant to its epistemic quality, important consequences follow. First of all, there is no obvious strategy of "maximizing" DIR or optimizing it in any other straightforward way. DIR is not a scalar quantity but a three-dimensional vector. What is more, the three dimensions of the vector are antagonistic to each other in the sense that each of them alone can easily be increased at the cost of one or both of the others, so that any methodological choice involves a trade-off between the three dimensions. Consequently, there can be no linear ranking of methods according to their purely epistemic quality. Since all three dimensions of DIR are geared towards the aim of getting at the truth, methodological choices therefore turn out to be underdetermined by truth as the aim of inquiry.

Another important consequence can be recognized when we turn to the question how the appropriate DIR for a particular case of inquiry can or should be determined. Is there a non-arbitrary way in which we may speak of the "correct" DIR? Early statisticians like Abraham Wald or C. West Churchman faced a similar question when they discussed the issue of how strong the statistical evidence must be in order to warrant the acceptance (or, respectively, the rejection) of a hypothesis. They were in agreement that the science of statistics cannot answer the question. Rather, the correct answer depends on what is at stake in the particular case (Wald 1942, 40–1; Churchman 1948, ch. 15). This insight about statistics can be generalized: a non-arbitrary sense of a correct DIR can only be determined as a function of certain value judgments.[12] The value judgments in question are judgments about *how* valuable, in the specific circumstances of each particular investigation, a correct positive result would be as compared to the state of continued ignorance, and similarly, how valuable a correct negative result would be, how bad a false negative result would be and how bad a false positive result would be. Without answers to these questions, meaningful comparisons between different distributions of inductive risks are underdetermined.[13]

---

[12] I explore the exact nature of this dependence in Wilholt (2013).

[13] This line of reasoning for the value-dependence of DIR is essentially the same as the one that Churchman's student Richard Rudner already used to argue that "The Scientist *qua* Scientist

5. Methodological Standards and the Distribution of Inductive Risks

While the only non-arbitrary sense of a correct DIR makes it dependent on value judgments, making the appropriate trade-offs and determining DIR for each and every investigation is in scientific practice not left to individual researchers and *their* value judgments. Instead, DIR is heavily constrained by the respective research community's methodological standards. As an illustration, consider again the methodological rule that you need a causal hypothesis in biosafety studies on GMO. By adopting this rule, you can lower the risk of getting false positive results (which in this case would mean affirming a positive correlation between the presence of the GMO and some undesirable event in the environment where in reality there is no causal connection between the two). For if your search for correlations is always guided by a causal hypothesis, you are less likely to be misled into positing some effect—concerning, say, the survival rates of one of the insect species in the environment—that later turns out to be an artefact of the study design or even a random pattern in your data-set. At the same time, the rule increases the risk of false negative results (i.e. the risk of missing an effect that really is there). This is so because studies that are *not* restricted by the proposed rule and that simply look for all sorts of changes that are statistically correlated with the presence of GMOs will arguably at least occasionally hit upon an environmental effect that is real but unexpected in the sense that we had no causal conjecture about it beforehand. (Both our causal understanding of biological mechanisms and our creativity in thinking about them are imperfect. Therefore, if it *seems* to us that our present understanding of the mechanisms provides no clue to a connection, it does not mean that there couldn't be one.) The proposed methodological rule would therefore go along with shifting some of the inductive risk from the false positives to the false negatives. To return to another one of our earlier examples, the selective publication of results of clinical

---

Makes Value Judgments," in his essay of the same title (1953). Owing in great part to the work of Heather Douglas (2000, 2009), who has rediscovered the argument's importance and explored its consequences and ramifications, it now again plays a central role in debates over science and values—rightly so, I believe (cf. Wilholt 2009, 2013). Note that my version of the argument in the present chapter does not start from particular assumptions about the moral responsibilities of scientists. Rather, the core argument is that there can be no linear measure of the quality of a method that derives only from the single aim of finding truth and avoiding falsehood. Methodological choices are therefore underdetermined by this aim. In *consequence*, every non-arbitrary methodological choice involves value judgments, and thus the (moral, social, political) responsibility of scientists is the only thing that can fill the gap which a "purely epistemic" perspective on methodology must necessarily leave.

trials has been suspected of leading to a situation in which the body of published research literature "overestimate[s] the benefits of an intervention" (Chan *et al.* 2004, 2457). Negative results regarding the efficacy of novel interventions (as compared to standard therapy) are particularly likely to fall under the carpet as they are often not published and are sometimes even cloaked in secrecy. The introduction of clinical trial repositories as a standard of the biomedical community makes it less likely that negative results fall into oblivion. Due to obligatory registration in publicly accessible repositories, negative outcomes too can become known to all researchers who take an interest in the respective studies, and this knowledge can then spread through the community. The rule thus limits the risk of false negative results, if under a "result" we choose in this case to understand the opinion that the research community will ultimately arrive at on the basis of the available information on concluded trials.

In face of the fact that methodological standards impose constraints on DIR, one could still attempt to uphold the view that the role of research communities in providing and maintaining such standards is merely accidental—auxiliary, as it were. In order to do so, one would have to argue that the standards save individual researchers the trouble of making the appropriate value judgments and finding the methodological choices that affect DIR accordingly by themselves. The obvious shortcoming of this line of reasoning is that it presupposes it to be somehow determined and uncontroversial what the appropriate value judgments are. While members of research communities tend to share *some* interests, it is unlikely that they would generally arrive at the same value judgments about the benefits of true results and the seriousness of mistakes. Methodological standards thus not only provide a service that makes the job of balancing out inductive risks easier. They also *harmonize* DIR within research communities. The value judgments implicit in the constraints on DIR that the standards set are in a sense binding for the community's members—not in the sense that the researchers are individually committed to endorsing the value judgments, but in the sense that they have to act as if they endorsed them when they perform certain research-related actions.

Should this imposition of a research community on its members perhaps be criticized or even opposed? On the contrary, I think it is hardly imaginable how a collective cognitive enterprise could work without it. In order to enable and maintain epistemic trust *within* a research collective, it is necessary to *co-ordinate* the DIR that is deemed acceptable in its research activities. At first sight it might seem a possible arrangement to let every scientist freely choose his or her own DIR in accordance with his or her own value judgments about how serious each particular kind of mistake would be. But such individual decisions would be extremely cumbersome to track and take account of by peers. Not only can value judgments vary considerably from individual to individual, it is also usually difficult to guess another person's value judgments on a given subject matter. But as we have seen, DIR is the decisive characteristic for assessing the trustworthiness of a given piece of scientific information for a given purpose. Methodological standards that put constraints on DIR thus make such an assessment of another researcher's results possible without having to take a guess at his or her value judgments.

That said, there is perhaps an additional reason why it is not at all objectionable if the value judgments that determine the limits on acceptable inductive risks are made by a collective rather than by each individual. Collectives or communities might simply be better placed to make such judgments, especially if they incorporate a diversity of perspectives and value outlooks. This may make them less prone to oversights and biases than an individual. While I agree that collective deliberation on value judgments can have these positive effects, my argument in this chapter does not rely on them. My main argument why constraints on DIR and the required implicit value judgment underlying them need to be determined on the community level is that harmonization is required in order to facilitate and maintain epistemic trust within research communities.

In conducting her scientific work as a member of a particular research community, a researcher implicitly commits herself to respecting its methodological standards and thereby to working within the limitations on acceptable distributions of inductive risks that these imply. In so far as this serves to harmonize DIR within the community, the precise constraints imposed by methodological standards turn out to be at least in part conventional in the following sense. The purpose of enabling trust within a community by harmonizing accepted distributions of inductive risks could presumably be served by a whole range of diverging determinations of what the ideal DIR is. With regard to the aim of facilitating reliable assessments of the trustworthiness of other researchers' results, it is crucial that everyone within the community sticks to *the same* standards and thus the same limitations on DIR, but not *which* particular DIR it is that is set as an ideal. The standards provide a solution to a problem of coordination and are thus conventions. They are only *partly* conventional, because they do not *only* serve the purpose of harmonizing DIR within the community. They also represent the research community's collective attempt to find the *right* balance between power and the two types of reliability. In that sense, they also represent an implicit consensus (or at least an implicit compromise position) of the community with regard to the question of how valuable the benefits of correct results and how grave the negative consequences of mistakes typically are for the kinds of research procedures that are subject to the standards at issue.

6. Conclusions

Research communities are bound together by methodological standards that shape DIR in their respective area of research. These standards serve as a solution to a problem of coordination and are thus irreducibly social—that is to say, communities play an *essential* role in shaping DIR. Without this role, the kind of collective epistemic enterprise that we call science, with its high degree of reliance on each others' results and its fast spread of novel information through the community, could arguably not get off the ground.

At the same time, DIR is decisive for assessing the trustworthiness of a piece of scientific information. It follows that every reasonable assessment of trustworthiness with regard to scientific results must attend to research communities. In placing our trust in a scientific result, we in part place our trust in the respective research community and its methodological standards. We trust that this community has set the limitations on the DIR in a suitable way, considering the risks and benefits of false and correct results appropriately. Ideally, someone investing trust in a scientific result would, of course, like to see the research communities weigh the risks and benefits in a way that is at least roughly in line with her own value judgments on the matter. (But note that a different, more circumspect kind of reliance in scientific results remains possible as long as the community determines the DIR in a *predictable* way, even if this diverges from how we ourselves as "users" of the information would have done it.) A part of our trust is, of course, also placed in the individual researcher (or collaborative group) generating the respective piece of scientific information.[14] A research project cannot be executed like an algorithm after all—for all the methodological conventions, skill and competence are of no less crucial importance in scientific research. But in the kind of collective enterprise that we call science, every individual researcher or collaborative group is also part of a research community whose standards impose decisive constraints on methodological choices and which therefore always has a share in the production of the results.

---

[14] I do therefore not wish to dispute that trust in individual colleagues still plays an important role in present-day science, as recent writers have emphasized (Shapin 2008; Frost-Arnold 2013). But the potential of trust directed at individuals alone to explain and serve as a basis of the division of cognitive labor in the sciences is very limited in light of the phenomena I have discussed. It has to be supplemented by trust directed at collective bodies.

Does the social diffusion of trustworthiness to the level of research communities bear any wider significance for the social epistemology of scientific research? For a start, it means that philosophers of science should take an interest in conventional methodological standards and the various ways they come about. Some standards take the form of tacit rules that encapsulate certain entrenched practices and are habitually passed on in professional training. Some originate as explicit recommendations of task forces or peer review panels initiated by professional associations or government agencies. Some are based on unilateral decisions by editors of influential journals. Some may even be inscribed in widely used research tools, such as statistics software packages. In investigating the standards and the ways they arise, philosophers should keep in mind that they inevitably contain implicit value judgments on the benefits of correct results and the disadvantages of mistakes.

The latter point leads to a more far-reaching conclusion. When we want to assess the relative merits of the collective methodological decisions of a research community, the aforementioned implicit value judgments will have to come under scrutiny too. How well do they reflect the epistemic role that the research community fulfils within wider society? This question cannot be answered with respect to just the single aim of getting at the truth. We have seen that it is a deeply problematic take on epistemic quality management to consider the reliability of a method to provide a simple, one-dimensional measure of how good a method is at getting at the truth. Power and both types of reliability are all aspects of aiming at the truth, but aspects that need to be traded off against one another. The question what the right balance between them is involves types of valuations that cannot themselves be derived from the aim of truth alone. Call an evaluative concept "purely epistemic" if it involves only evaluations with regard to the aim of acquiring true beliefs and avoiding false ones. Then what we have observed means that there can be no purely epistemic linear measure of the quality of a method. A purely epistemic perspective on methodology is always incomplete.[15]

---

[15] This chapter was first presented at the March 2011 conference on Collective Epistemology, organized by Miranda Fricker and Michael Brady, from which this volume originates. Different parts of it were also presented at the SPSP conference in Exeter in June 2011, at the EPSA in Athens in October 2011, at a conference on The Social Relevance of Philosophy of Science at the Center for Interdisciplinary Research in Bielefeld in June 2012 and at the GAP.8 in Konstanz in September 2012. I am grateful to the organizers and participants of all these events, and particularly to my co-panelists at Exeter and Athens: Justin Biddle, Bryce Huebner, Rebecca Kukla, and Eric Winsberg.

Bibliography

Adams, Robert Merrihew (1985). 'Involuntary Sins', *Philosophical Review*, 94(1): 3–31.
Alchourrón, Carlos, Peter Gärdenfors, and David Makinson (1985). 'On the Logic of Theory Change: Partial Meet Contraction and Revision Functions', *Journal of Symbolic Logic*, 50: 510–30.
Allport, Gordon (1954). *The Nature of Prejudice*. Cambridge, MA: Addison-Wesley.
Anderson, Elizabeth (2007). 'Fair Opportunity in Education: A Democratic Equality Perspective', *Ethics*, 117: 595–622.
Anderson, Elizabeth (2010). *The Imperative of Integration*. Princeton: Princeton University Press.
Annas, Julia (2011). *Intelligent Virtue*. Oxford: Oxford University Press.
Anscombe, G. E. M. (1957). *Intention*. Cambridge, MA: Harvard University Press.
Armstrong, D. H. (1981). *The Nature of Mind and Other Essays*. Ithaca, NY: Cornell University Press.
The ATLAS Collaboration, G. Aad, E. Abat *et al.* (2008). "The ATLAS Experiment at the CERN Large Hadron Collider", *Journal of Instrumentation* 3, S08003.
Bach, Kent (2008). 'Applying Pragmatics to Epistemology', *Philosophical Issues*, 18: 68–88.
Bach, Kent and Robert Harnisch (1979). *Linguistic Communication and Speech Acts*. Cambridge, MA: MIT Press.
Baehr, Jason (2011). *The Inquiring Mind: On Intellectual Virtues and Virtue Epistemology*. Oxford: Oxford University Press.
Barrett, H. Clark (2005). 'Adaptations to Predators and Prey'. In *The Handbook of Evolutionary Psychology*, ed. David Buss. New York: John Wiley & Sons.
Barry, Christian (2005). 'Applying the Contribution Principle'. In *Global Institutions and Responsibilities: Achieving Global Justice*, ed. C. Barry and T. Pogge. Malden, MA, and Oxford: Blackwell: 280–97.
Beatty, John (2006). 'Masking Disagreement among Experts', *Episteme*, 3(1): 52–67.
Bénabou, Roland and Jean Tirole (2011). 'Identity, Morals, and Taboos: Beliefs as Assets', *Quarterly Journal of Economics*, 126(2): 805–55.
Bertrand, Marianne and Sendhil Mullainathan (2004). 'Are Emily and Greg More Employable than Lakisha and Jamal? A Field Experiement on Labor Market Discrimination', *American Economic Review*, 94(4): 991–1013.
Bicchieri, Christina (2006). *The Grammar of Society: The Nature and Dynamics of Social Norms*. Cambridge: Cambridge University Press.
Bicchieri, Cristina and Alex K. Chavez (2013). 'Norm Manipulation, Norm Evasion: Experimental Evidence', *Economics and Philosophy*, 29(2): 175–98.
Bird, Alexander (2010). 'Social Knowing: The Social Sense of "Scientific Knowledge"', *Philosophical Perspectives*, 24: 23–56.
Borgerson, Kirstin (2009). 'Valuing Evidence: Bias and the Evidence Hierarchy of Evidence-Based Medicine', *Perspectives in Biology and Medicine*, 52(2): 218–33.

Bouvier, Alban (2004). 'Individual Beliefs and Collective Beliefs in Sciences and Philosophy: The Plural Subject and the Polyphonic Subject Accounts', *Philosophy of the Social Sciences*, 34(3): 382–407.
Brady, Michael (2013). *Emotional Insight: The Epistemic Role of Emotional Experience*. Oxford: Oxford University Press.
Bratman, Michael (1999). *Faces of Intention: Selected Essays on Intention and Agency*. Cambridge: Cambridge University Press.
Broer, Inge (2012). 'Divergierende naturwissenschaftliche Bewertung der Grünen Gentechnik: Grundlagen biologischer Risikoanalyse'. In *Grüne Gentechnik: Zwischen Forschungsfreiheit und Anwendungsrisiko*, ed. Herwig Grimm and Stephan Schleissing. Baden Baden: Nomos: 81–92.
Broer, Inge and Joachim, Schiemann (2009). 'Biologische Sicherheitsforschung an gentechnisch veränderten Pflanzen', discussion paper.
Brown, Joseph (1865). *Southern Watchman*, 1 March: <http://athnewspapers.galileo.usg.edu/athnewspapers/view?docId=news/swm1865/swm1865-0033.xml>.
Buchanan, Allen (2002). 'Social Moral Epistemology', *Social Philosophy and Policy*, 19(2): 126–52.
Buechner, Frederick (1994). 'A Sprig of Hope'. In *A Chorus of Witnesses: Model Sermons for Today's Preacher*, ed. Thomas Long and Cornelius Plantinga, Jr. Grand Rapids, MI: Eerdmans.
Camerer, Colin (2003). *Behavioral Game Theory: Experiments in Strategic Interaction*. New York: Russell Sage Foundation.
Capelli, Andrea, Elena Castellani, Filippo Colomo, and Paolo Di Vecchia (2012). *The Birth of String Theory*. New York: Cambridge University Press.
Cappelen, Alexander W., Astri Drange Hole, Erik Ø. Sørensen, and Bertil Tungodden (2011). 'The Importance of Moral Reflection and Self-reported Data in a Dictator Game with Production', *Social Choice and Welfare*, 36(1): 105–20.
Carnap, Rudolf (1956). *Meaning and Necessity: A Study in Semantics and Modal Logic*. Chicago: University of Chicago Press.
Carroll, Lewis (1895). 'What the Tortoise said to Achilles', *Mind*, 4: 278–80.
Chan, An-Wen, Asbjørn Hróbjartsson, Mette T. Haahr, Peter C. Gøtzsche, and Douglas G. Altman (2004). 'Empirical Evidence for Selective Reporting of Outcomes in Randomized Trials: Comparison of Protocols to Published Articles', *Journal of the American Medical Association*, 291: 2457–65.
Chignell, Andrew (2013). 'The Ethics of Belief'. In *The Stanford Encyclopedia of Philosophy*, ed. Edward N. Zalta. Spring 2013 edn. <plato.stanford.edu/archives/spr2013/entries/ethics-belief>.
Christensen, David (2007). 'Epistemology of Disagreement: The Good News', *Philosophical Review*, 116: 187–217.
Christensen, David (2011). 'Disagreement, Question-Begging, and Epistemic Self-Criticism', *Philosopher's Imprint*, 11: <http://hdl.handle.net/2027/spo.3521354.0011.006>.
Churchman, C. West (1948). *Theory of Experimental Inference*. New York: Macmillan.
Cialdini, Robert (2001). 'The Science of Persuasion', *Scientific American*, 284: 76–81.
Clark, Lee Anna and David Watson (1994). 'Distinguishing Functional from Dysfunctional Affective Responses'. In Ekman and Davidson (1994).

Cobb, Howell (1865). Letter to J. A. Seddon, Secretary of War of the Confederate States of America. Macon, Ga., 8 January. Published in *The American Historical Review*, 1(1) (Oct. 1895): 97–8.

Cole, K. C. (1987). 'A Theory of Everything', *New York Times Magazine*, 18 October, at: <http://www.nytimes.com/1987/10/18/magazine/a-theory-of-everything.html>.

Collins, Stephanie (2013). 'Collectives' Duties and Collectivization Duties', *Australasian Journal of Philosophy*, 91(2): 231–48.

Condorcet, Nicolas de. (1781). *Réflexions sur l'esclavage des nègres*. Neufchatel: Société typographique.

Condorcet, Nicolas de. ([1790] 1912). *On the Admission of Women to the Rights of Citizenship*, ed. Alice Vickery. Letchworth: Garden City Press: <http://oll.libertyfund.org/?option=com_staticxt&staticfile=show.php%3Ftitle=1013&chapter=81939&layout=html&Itemid=27>.

Cottingham, John (2003). *On the Meaning of Life*. New York: Routledge.

Czopp, Alexander M., Margo J. Monteith, and Aimee Y. Mark (2006). 'Standing Up for a Change: Reducing Bias through Interpersonal Confrontation', *Journal of Personality and Social Psychology*, 90(5): 784–803.

Dagger, Richard (2007). 'Political Obligation'. In *The Stanford Encyclopedia of Philosophy* (17 Apr. 2007; rev. 30 Apr. 2010). Available at: <http://plato.stanford.edu/entries/political-obligation/>.

Dana, Jason, Roberto A. Weber, and Jason Xi Kuang (2007). 'Exploiting Moral Wiggle Room: Experiments Demonstrating an Illusory Preference for Fairness', *Economic Theory*, 33(1): 67–80.

D'Arms, Justin (2005). 'Two Arguments for Sentimentalism', *Philosophical Issues*, 15(1): 1–21.

D'Arms, Justin and Daniel Jacobson (2006). 'Anthropocentric Constraints on Human Value'. In *Oxford Studies in Metaethics* 1, ed. Russ Shafer-Landau. Oxford: Oxford University Press.

Darwall, Stephen (2006). *The Second Person Standpoint*. Cambridge: Cambridge University Press.

Dawid, Richard (2013). *String Theory and the Scientific Method*. New York: Cambridge University Press.

DeRose, K. (1996). 'Knowledge, Assertion, and Lotteries', *Australasian Journal of Philosophy*, 74: 568–80.

DeRose, Keith (2000). 'Solving the Skeptical Problem'. In *Epistemology: An Anthology*, ed. Ernest Sosa and Jaegwon Kim. Oxford: Blackwell: 482–502.

de Sousa, Ronald (1988). *The Rationality of Emotion*. Cambridge, MA: MIT Press.

Dewey, John (1922). *Human Nature and Conduct: An Introduction to Social Psychology*. New York: H. Holt & Co.

Dewey, John and James Tufts ([1932] 1981). *Ethics*, ed. Jo Ann Boydston. *The Later Works, 1925–1953*. Carbondale, Ill.: Southern Illinois University Press.

Dietrich, Franz and Kai Spiekermann (2013). 'Epistemic Democracy with Defensible Premises', *Economics and Philosophy*, 29(1): 87–120.

Douglas, Heather (2000). 'Inductive Risk and Values in Science', *Philosophy of Science*, 67: 559–79.

Douglas, Heather (2009). *Science, Policy, and the Value-Free Ideal*. Pittsburgh: University of Pittsburgh Press.
Dubois, Laurent (2004). *Avengers of the New World: The Story of the Haitian Revolution*. Cambridge, MA: Belknap/Harvard University Press.
Dubois, Laurent (2006). 'An Enslaved Enlightenment: Rethinking the Intellectual History of the Atlantic', *Social History*, 31(1): 1–14.
Dubois, Laurent (2012). *Haiti: The Aftershocks of History*. New York: Metropolitan Books.
Dubois, Laurent and John D. Garrigus (2006). *Slave Revolution in the Caribbean, 1789–1804: A Brief History with Documents*. Basingstoke and New York: Palgrave Macmillan.
Ekman, Paul and Richard Davidson (1994). *The Nature of Emotion*. Oxford: Oxford University Press.
Elga, Adam (2010). 'How to Disagree about How to Disagree'. In *Disagreement*, ed. Richard Feldman and Ted A. Warfield. Oxford: Oxford University Press: 175–86.
Ellsworth, Phoebe (1994). 'Levels of Thought and Levels of Emotion'. In Ekman and Davidson.
Enoch, David (2012). 'Being Responsible, Taking Responsibility, and Penumbral Agency'. In *Luck, Value, and Commitment: Themes from the Ethics of Bernard Williams*, ed. Ulrike Heuer and Gerald Lang. Oxford: Oxford University Press.
Erskine, Toni (2001). 'Assigning Responsibilities to Institutional Moral Agents: The Case of States and Quasi-states', *Ethics and International Affairs*, 15(2): 67–89.
Estlund, David (2008). *Democratic Authority*. Princeton: Princeton University Press.
Fagan, Melinda Bonnie (2011). 'Is There Collective Scientific Knowledge? Arguments from Explanation', *Philosophical Quarterly*, 61(243): 247–69.
Fagan, Melinda Bonnie (forthcoming *a*). 'Do Groups Have Scientific Knowledge?' In *From Individual to Collective Intentionality*, ed. Sara Chant, Frank Hindriks, and Gerhard Preyer. New York: Oxford University Press.
Fagan, Melinda Bonnie (forthcoming *b*). 'Collective Scientific Knowledge', *Philosophy Compass*.
Feinberg, Joel (1968). 'Collective Responsibility', *Journal of Philosophy*, 65(21): 674–88.
Feldman, Richard (2000). 'The Ethics of Belief', *Philosophy and Phenomenological Research*, 60(3): 667–95.
Festinger, Leon and James M. Carlsmith (1959). 'Cognitive Consequences of Forced Compliance', *Journal of Abnormal and Social Psychology*, 58(2): 203–10.
Fetterolf, Elianna (2014). 'Remorse: A Prospective Genealogy'. Unpublished PhD thesis, University of Sheffield.
Financial Times (2013). 'Dublin Plans Anglo Irish Inquiry Amid Public Anger', 25 June.
Fiske, Susan (1998). 'Stereotyping, Prejudice and Discrimination'. In *Handbook of Social Psychology*, Vol. 2, ed. Daniel Gilbert, Susan Fiske, and Gardner Lindzey. New York: McGraw-Hill.
Foner, Eric (1995). *Free Soil, Free Labor, Free Men: The Ideology of the Republican Party Before the Civil War*. New York: Oxford University Press.
Foner, Eric (2007). *Nothing but Freedom: Emancipation and Its Legacy*. Baton Rouge, LA: Louisiana State University Press.

Francis, Robert (QC), Chair (2003). *Report of the Mid Staffordshire NHS Foundation Trust Public Inquiry*, Volume 1: *Analysis of evidence and lessons learned (part 1)*. London: The Stationary Office.
Fricker, Miranda (1998). 'Rational Authority and Social Power: Towards a Truly Social Epistemology', *Proceedings of the Aristotelian Society*, 98: 159–77.
Fricker, Miranda (2007). *Epistemic Injustice: Power and the Ethics of Knowing*. Oxford: Oxford University Press.
Fricker, Miranda (2010). 'Can There Be Institutional Virtues?' In *Oxford Studies in Epistemology (Special Theme: Social Epistemology)*, Vol. 3, ed. Tamar Szabó Gendler and John Hawthorne. Oxford: Oxford University Press: 235–52.
Fricker, Miranda (2012). 'The Relativism of Blame and Williams's Relativism of Distance', *Proceedings of the Aristotelian Society*, supp. vol. 84: 151–77.
Fricker, Miranda (2014). 'What's the Point of Blame? A Paradigm Based Explanation', *Noûs*. DOI: 10.1111/nous.12067.
Friedman, Michael (2001). *Dynamics of Reason*. Chicago: University of Chicago Press.
Frost-Arnold, Karen (2013). 'Moral Trust and Scientific Collaboration', *Studies in History and Philosophy of Science*, 44 (3): 301–10.
Gadamer, Hans-Georg (1989). *Truth and Method*, 2nd rev. edn., trans. Joel Weinsheimer and Donald Marshall. New York: Crossroad.
Gaita, Raimond (1998). *A Common Humanity: Thinking About Love and Truth and Justice*. London: Routledge.
Gaita, Raimond (2011). *After Romulus*. Melbourne: Text Publishing.
Galison, Peter (1995). 'Theory Bound and Unbound: Superstrings and Experiments'. In *Laws of Nature: Essays on the Philosophical, Scientific and Historical Dimensions*, ed. Friedel Weinert. Berlin: de Gruyter: 369–408.
Gallie, W. B. (1955-6). 'Essentially Contested Concepts', *Proceedings of the Aristotelian Society*, NS 56: 167–98.
Galton, Francis (1907). 'Vox Populi (the Wisdom of Crowds)', *Nature*, 75: 450–1.
Gärdenfors, Peter (2008). *Knowledge in Flux*. (1st edn. 1988). London: College Publications.
Gaus, Gerald (2011). 'On Seeking the Truth (Whatever That Is) through Democracy: Estlund's Case for the Qualified Epistemic Claim', *Ethics*, 121: 270–300.
Gendler, Tamar Szabó (2014). 'The Third Horse: On Unendorsed Association and Human Behaviour', *Proceedings of the Aristotelian Society*, supp. vol. 38: 185–218.
Gettier, Edmund (1963). 'Is Justified True Belief Knowledge?', *Analysis*, 23: 121–3.
Gibbard, Allan (2002). 'Knowing What To Do, Seeing What To Do'. In *Ethical Intuitionism: Re-evaluations*, ed. Philip Stratton-Lake. Oxford: Oxford University Press: 212–28.
Giere, Ronald N. and Barton Moffatt (2003). 'Distributed Cognition: Where the Cognitive and the Social Merge', *Social Studies of Science*, 33(2): 301–10.
Gilbert, Margaret (1987). 'Modeling Collective Belief', *Synthese*, 73: 185–204.
Gilbert, Margaret (1989). *On Social Facts*. Princeton: Princeton University Press).
Gilbert, Margaret (1993). 'Group Membership and Political Obligation', *The Monist*, 76: 119–31.
Gilbert, Margaret (1996). *Living Together*. Lanham, MD: Rowman & Littlefield.
Gilbert, Margaret (2000a). 'Collective Belief and Scientific Change'. In Gilbert (2000b).

Gilbert, Margaret (2000b). *Sociality and Responsibility: New Essays in Plural Subject Theory*. Lanham, MD: Rowman & Littlefield.
Gilbert, Margaret (2000c). 'What Is It for *Us* to Intend?' In Gilbert (2000b), 14–36.
Gilbert, Margaret (2001a). 'Collective Guilt and Collective Guilt Feelings', *Journal of Ethics*, 6: 115–43.
Gilbert, Margaret (2001b). 'Collective Preferences, Obligations and Rational Choice', *Economics and Philosophy*, 17(1): 109–19.
Gilbert, Margaret (2002). 'Belief and Acceptance as Features of Groups', *Protosociology*, 16: 35–69.
Gilbert, Margaret (2004). 'Collective Epistemology', *Episteme*, 1(2): 95–107.
Gilbert, Margaret (2012). 'Giving Claim-Rights Their Due'. In *Rights: Concepts and Contexts*, ed. Brian Bix and Horacio Spector. Farnham: Ashgate: 301–23.
Gilbert, Margaret (2013). *Joint Commitment: How We Make the Social World*. Oxford: Oxford University Press.
Gilbert, Margaret and Daniel Pilchman (2014). 'Belief, Acceptance, and What Happens in Groups: Some Methodological Considerations'. In *Essays in Collective Epistemology*, ed. Jennifer Lackey. Oxford: Oxford University Press.
Godfrey-Smith, Peter (2003). *Theory and Reality: An Introduction to Philosophy of Science*. Chicago: University of Chicago Press.
Goldberg, Sanford C. (2006). 'The Social Diffusion of Warrant and Rationality', *Southern Journal of Philosophy*, 44: 118–38.
Goldberg, S. (2013a). 'Disagreement, Defeaters, and Assertion'. In *The Epistemology of Disagreement*, ed. David Christensen and Jennifer Lackey. Oxford: Oxford University Press: 167–89.
Goldberg, S. (2013b). 'Defending Philosophy in the Face of Systematic Disagreement'. In *Disagreement and Skepticism*, ed. Diego Machuca. New York: Routledge: 277–94.
Goldenberg, Maya J. (2006). 'On Evidence and Evidence-Based Medicine: Lessons from the Philosophy of Science', *Social Science and Medicine*, 62(11): 2621–32.
Goldman, Alvin I. and Whitcomb, Dennis, eds. (2010). *Social Epistemology: An Anthology*. Oxford: Oxford University Press.
Goldman, Alvin I. (1992). *Liaisons: Philosophy Meets the Cognitive and Social Sciences*. Cambridge, MA: MIT Press.
Goodin, Robert (2007). 'Enfranchising All Affected Interests, and its Alternatives', *Philosophy and Public Affairs*, 35(1): 40–68.
Goodin, Robert E. (1995). *Utilitarianism as a Public Philosophy*. Cambridge: Cambridge University Press.
Goodin, Robert E. (2009). 'Demandingness as a Virtue', *Journal of Ethics*, 13: 1–13.
Grasswick, Heidi (2013). 'Feminist Social Epistemology'. In *The Stanford Encyclopedia of Philosophy*, ed. Edward N. Zalta. Spring 2013 edn.: <http://plato.stanford.edu/archives/spr2013/entries/feminist-social-epistemology/>.
Grasswick, Heidi E. and Webb, Mark Owen (2002). 'Feminist Epistemology as Social Epistemology', *Social Epistemology: A Journal of Knowledge, Culture and Policy*, 16(3): 185–96.
Greene, Brian (1999). *The Elegant Universe: Superstrings, Hidden Dimensions, and the Quest for the Ultimate Theory*. New York: W.W. Norton & Company.

Grice, P. (1968/1989). 'Logic and Conversation'. Reprinted in *Studies in the Way of Words*. Cambridge, MA: Harvard University Press.
Grimm, Stephen (2006). 'Is Understanding a Species of Knowledge?', *British Journal for the Philosophy of Science*, 57: 515–35.
Grossman, Zachary and Joel van der Weele (forthcoming). 'Self-Image and Wilful Ignorance in Social Decisions', *Journal of the European Economic Association*.
The GUSTO Investigators (1993). 'An International Randomized Trial Comparing Four Thrombolytic Strategies for Acute Myocardial Infarction', *New England Journal of Medicine*, 329: 673–82.
Güth, Werner and Steffen Huck (1997). 'From Ultimatum Bargaining to Dictatorship— An Experimental Study of Four Games Varying in Veto Power', *Metroeconomica*, 48(3): 262–99.
Güth, Werner, Steffen Huck, and Peter Ockenfels (1996). 'Two-Level Ultimatum Bargaining with Incomplete Information: An Experimental Study', *Economic Journal*, 106 (436): 593–604.
Haddock, Adrian, Alan Millar, and Duncan Pritchard, eds. (2010). *Social Epistemology*. Oxford: Oxford University Press.
Haisley, Emily C. and Roberto A. Weber (2010). 'Self-serving Interpretations of Ambiguity in Other-regarding Behavior', *Games and Economic Behavior*, 68(2): 614–25.
Hall, Richard J. and Charles R. Johnson (1998). 'The Epistemic Duty to Seek More Evidence', *American Philosophical Quarterly*, 35(2): 129–39.
Hardwig, John (1985). 'Epistemic Dependence', *Journal of Philosophy*, 82(7): 335–49.
Hatfield, Elaine, John Cacioppo, and Richard L. Rapson (1994). *Emotional Contagion*. New York: Cambridge University Press.
Hawthorne, John (2003). *Knowledge and Lotteries*. Oxford: Oxford University Press.
Held, Virginia (1970). 'Can a Random Collective of Individuals be Morally Responsible?', *Journal of Philosophy*, 68(14): 471–81.
Hempel, Carl G. (1965). *Aspects of Scientific Explanation*. New York: Free Press.
Herzog, Don (1989). *Happy Slaves*. Chicago: University of Chicago Press.
Hobbes, Thomas ([1651] 1994). *Leviathan*. Indiana: Hackett.
Holroyd, Jules (2012). 'Responsibility for Implicit Bias', *Social Philosophy*, 43(3) Fall: 274–306.
Holt, Thomas (2000). 'The Essence of the Contract: The Articulation of Race, Gender, and Political Economy in British Emancipation Policy, 1838–1866'. In *Beyond Slavery*, ed. Frederick Cooper, Thomas C. Holt, and Rebecca J. Scott. Chapel Hill: University of North Carolina Press: ch. 1.
Hume, David ([1739] 1978). *A Treatise of Human Nature*. 2nd edn., ed. L. A. Selby-Bigge and P. H. Nidditch. Oxford: Clarendon Press.
International Committee of Medical Journal Editors (ICMJE) (2010). 'Toward More Uniform Conflict Disclosures.' *Editorial*.
International Committee of Medical Journal Editors (ICMJE) (2014). 'Recommendations for the Conduct, Reporting, Editing, and Publication of Scholarly Work in Medical Journals', updated December 2014, <http://www.icmje.org/recommendations/> (accessed 9/1/2015).

James, C. L. R. (1963). *The Black Jacobins: Toussaint L'Ouverture and the San Domingo Revolution*. 2nd edition, revised. New York: Vintage Books.

Jasentuliyana, Nandasiri (1995). Introduction. In *Perspectives on International Law*, ed. Nandasiri Jasentuliyana. The Hague: Martinus Nijhoff Publishers.

Jefferson, Thomas ([1785] 1905). 'Notes on Virginia II'. In *The Works of Thomas Jefferson*, ed. Paul Leicester Ford. New York: G. P. Putnam's Sons.

Johnson, George (1998). 'Almost in Awe, Physicists Ponder "Ultimate" Theory', *New York Times*, 22 September. Archived at: <www.nytimes.com/1998/09/22/science/almost-in-awe-physicists-ponder-ultimate-theory.html>.

Kelly, Thomas (2005). 'The Epistemic Significance of Disagreement'. In *Oxford Studies in Epistemology*, Vol. 1, ed. Tamar Szabó Gendler and John Hawthorne: 1167–96.

Kelly, Thomas (2010). 'Peer Disagreement and Higher Order Evidence'. In *Disagreement*, ed. Richard Feldman and Ted A. Warfield. Oxford: Oxford University Press: 111–74.

Kelsen, Hans (1955). 'Foundations of Democracy', *Ethics*, 66: 1–101.

Kind, Amy (2003). 'Shoemaker, Self-Blindness and Moore's Paradox', *Philosophical Quarterly*, 53: 39–48.

Kitcher, Philip (1994). 'Contrasting Conceptions of Social Epistemology'. In *Socializing Epistemology: The Social Dimensions of Knowledge*, ed. Frederick F. Schmitt. Lanham, MD: Rowman & Littlefield: 111–34.

Knorr-Cetina, Karin (1999). *Epistemic Cultures: How the Sciences Make Knowledge*. Cambridge, MA: Harvard University Press.

Konow, James (2000). 'Fair Shares: Accountability and Cognitive Dissonance in Allocation Decisions', *American Economic Review*, 90(4): 1072–91.

Kornblith, Hilary (1983). 'Justified Belief and Epistemically Responsible Action', *Philosophical Review*, 92(1): 33–48.

Kuhn, Thomas (1962). *The Structure of Scientific Revolutions*. Chicago: University of Chicago Press.

Kukla, Rebecca (2012). '"Author TBD": Radical Collaboration in Contemporary Biomedical Research", *Philosophy of Science*, 79(5): 845–58.

Kvanvig, Jonathan (1992). *The Intellectual Virtues and the Life of the Mind*. Lanham, MD: Rowman & Littlefield.

Lackey, Jennifer, ed. (2014). *Essays in Collective Epistemology*. Oxford: Oxford University Press.

Lakatos, Imre (1970). 'Falsification and the Methodology of Scientific Research'. In *Criticism and the Growth of Knowledge*, ed. Imre Lakatos and Alan Musgrave. New York: Cambridge University Press, 91–195.

Larmore, Charles (2008). *The Autonomy of Morality*. Cambridge: Cambridge University Press.

Laudan, Larry (1977). *Progress and its Problems: Toward a Theory of Scientific Growth*. Berkeley: University of California Press.

Laurence, Ben (2011). 'An Anscombian Approach to Collective Action'. In *Essays on Anscombe's 'Intention'*, ed. A. Ford, J. Hornsby, and F. Stoutland. Cambridge, MA: Harvard University Press: 270–94.

Lawford-Smith, Holly (2012). 'The Feasibility of Collectives' Actions', *Australasian Journal of Philosophy*, 90(3): 453–67.

Lazarus, Richard (1994). 'Appraisal: The Long and the Short of It'. In Ekman and Davidson.
LeDoux, Joseph (1996). *The Emotional Brain*. New York: Simon & Schuster.
Leith, John, ed. (1982). *Creeds of the Churches*, 3rd edn. Louisville, KY: John Knox Press.
Leslie, Sara-Jane (forthcoming). 'The Original Sin of Cognition: Fear, Prejudice and Generalization', *Journal of Philosophy*.
Lewis, David (1969). *Convention: A Philosophical Study*. Oxford: Wiley-Blackwell.
List, Christian and Robert Goodin (2001). 'Epistemic Democracy: Generalizing the Condorcet Jury Theorem', *Journal of Political Philosophy*, 9: 277–306.
List, Christian and Philip Pettit (2011). *Group Agency: The Possibility, Design, and Status of Corporate Agents*. Oxford: Oxford University Press.
Locke, John ([1689] 2004). *Second Treatise of Government*. New York: Barnes & Noble.
Loetscher, Lefferts (1983). *A Brief History of the Presbyterians*, 4th edn. Philadelphia: Westminster.
Longino, Helen (1990). *Science as Social Knowledge: Values and Objectivity in Scientific Inquiry*. Princeton: Princeton University Press.
Longino, Helen (2002). *The Fate of Knowledge*. Princeton: Princeton University Press.
Lord Laming (2004). 'Memorandum at the Select Committee on Public Administration, Minutes of Evidence, for the Victoria Climbié Inquiry': <http://www.publications.parliament.uk/pa/cm200304/cmselect/cmpubadm/>.
Louis XIV (1685). 'Code Noir': <http://chnm.gmu.edu/revolution/d/335/>. Accessed 29/8/2012.
MacIntyre, Alasdair (1983). *After Virtue*, 2nd edn. Notre Dame, IN: University of Notre Dame Press.
Madva, Alex (unpublished manuscript). 'Biased Against De-biasing: On the Role of (Institutionally Sponsored) Self-transformation in the Struggle against Prejudice'.
Maitra, Ishani (2010). 'The Nature of Epistemic Injustice', *Philosophical Books*, 51(4): 196–211.
Mathiesen, Kay (2006). 'The Epistemic Features of Group Belief', *Episteme*, 2(3): 161–75.
Mathieson, Kay, ed. (2007). *Collective Knowledge and Collective Knowers*, Special Issue of *Social Epistemology: A Journal of Knowledge, Culture and Policy*, 21(3).
May, Larry (2006). 'State Aggression, Collective Liability, and Individual Mens Rea', *Midwest Studies in Philosophy*, 30: 309–24.
McFarlane, J. (2011). 'What Is Assertion?' In *Assertion: New Essays*, ed. Jessica Brown and Herman Cappelen. Oxford: Oxford University Press: 79–96.
Medina, José (2013). *The Epistemology of Resistance: Gender and Racial Oppression, Epistemic Injustice, and Resistant Imaginations*. Oxford and New York: Oxford University Press.
Merali, Zeeya (2011). 'Collaborative Physics: String Theory Finds a Bench Mate', *Nature*, 478: 302–4.
Miller, David (2009). 'Democracy's Domain', *Philosophy and Public Affairs*, 37(3): 201–28.
Mills, Charles (2007). 'White Ignorance'. In *Race and Epistemologies of Ignorance*, ed. Shannon Sullivan and Nancy Tuana. Albany, NY: State University of New York Press: 11–38.

Mitzkewitz, Michael and Rosemarie Nagel (1993). 'Experimental Results on Ultimatum Games with Incomplete Information', *International Journal of Game Theory*, 22(2): 171–98.

Moran, Richard (2001). *Authority and Estrangement: An Essay on Self-Knowledge*. Princeton: Princeton University Press.

Moss-Racusin, Corinne A., John F. Dovidio, Victoria L. Brescoll, Mark J. Graham, and Jo Handelsman (2012). 'Science Faculty's Subtle Gender Biases Favor Male Students', *Proceedings of the National Academy of Sciences of the United States of America*, 109(41): 16474–9.

Myers, Benjamin (2009). 'Disruptive History: Rowan Williams on Heresy and Orthodoxy'. In *On Rowan Williams: Critical Essays*, ed. Matheson Russell. Eugene, OR: Cascade Books: 47–67.

Nagel, Jennifer (2014). 'Intuition, Reflection, and the Command of Knowledge', *Proceedings of the Aristotelian Society*, supp. vol. 38: 219–41.

Nall, Jeff (2008). 'Condorcet's Personal Relationships and Why He Became a Feminist' (download from Academia.edu).

National Science Board (2012). *Science and Engineering Indicators 2012*. Arlington: National Science Foundation: <http://www.nsf.gov/statistics/seind12/pdf/seind12.pdf> (accessed 25/3/2015).

NTP (2001). *National Toxicology Program's report of the endocrine disruptors low-dose peer review*. National Toxicology Program, U.S. Department of Health and Human Services: <http://ntp.niehs.nih.gov/ntp/htdocs/liason/LowDosePeerFinalRpt.pdf> (accessed 25/3/2015).

Oakes, James (1990). *Slavery and Freedom: An Interpretation of the Old South*. New York: Alfred A. Knopf.

O'Brien, Lilian (2003). Review of Richard Moran's *Authority and Estrangement*, *European Journal of Philosophy*, 11: 375–90.

Ockenfels, Axel and Peter Werner (2012). 'Hiding Behind a Small Cake in a Newspaper Dictator Game', *Journal of Economic Behavior and Organization*, 82(1): 82–5.

O'Donovan, Oliver (2008). *Church in Crisis*. Eugene, OR: Cascade Books.

Pagin, P. (2011). 'Information and Assertoric Force'. In *Assertion: New Essays*, ed. Jessica Brown and Herman Cappelen. Oxford: Oxford University Press: 97–136.

Parrish, John M. (2009). 'Collective Responsibility and the State', *International Theory*, 1(1): 119–54.

Pasternak, Avia (2013). 'Limiting States' Corporate Responsibility', *Journal of Political Philosophy*, 21(4): 361–81.

Paul, Sarah K. (2012). 'How We Know What We Intend', *Philosophical Studies*, 161: 327–46.

Peacocke, Cristopher (2003). 'Implicit Conceptions, Understanding, and Rationality'. In *Reflections and Replies: Essays on the Philosophy of Tyler Burge*, ed. Tyler Burge, Martin Hahn, and Bjørn T. Ramberg. Cambridge, MA: MIT Press: 117–52.

Peter, Fabienne (2009). *Democratic Legitimacy*. New York: Routledge.

Peter, Fabienne (2013). 'The Procedural Epistemic Value of Deliberation', *Synthese*, 190(7): 1253–66.

Pettigrove, Glen and Nigel Parsons (2012). 'Shame: A Case Study of Collective Emotion', *Social Theory and Practice*, 38: 504–30.

Pettit, Philip (2007). 'Rationality, Reasoning, and Group Agency', *Dialectica* 61: 495–519.
Pettit, Philip (2010). 'Groups with Minds of Their Own'. In *Social Epistemology: Essential Readings*, ed. Alvin I. Goldman and Dennis Whitcomb. Oxford: Oxford University Press.
Pettit, Philip and David Schweikard (2006). 'Joint Actions and Group Agents', *Philosophy of the Social Sciences*, 36: 18–39.
Presbyterian Church (USA) (1999). *The Book of Confessions*. Louisville, KY: Office of the General Assembly of the Presbyterian Church (USA).
Presbyterian Church (USA) (2013). *The Book of Order*, 2013–2015. Louisville, KY: Office of the General Assembly of the Presbyterian Church (USA).
Putnam, Hilary (1978). *Meaning and the Moral Sciences*. London: Routledge & Kegan Paul.
Quinton, Anthony (1975/6). 'Social Objects', *Proceedings of the Aristotelian Society*, 76: 1–27.
Rawls, John (1971). *A Theory of Justice*. Cambridge, MA: Harvard University Press.
Rawls, John (1999). 'The Idea of an Overlapping Consensus'. In *John Rawls: Collected Papers*, ed. Samuel Freeman. Cambridge, MA: Harvard University Press: 421–48.
Reid, Thomas ([1788] 1969). *Essays on the Active Powers of the Human Mind*, ed. Baruch Brody. Cambridge, MA: MIT Press.
Reidy, David (2004). 'Rawls on International Justice: A Defense', *Political Theory*, 32: 291–319.
Riach, Peter A. and Judith Rich (2002). 'Field Experiments of Discrimination in the Market Place', *Economic Journal*, 112(483): F480–518.
Ricoeur, Paul (1988). *Time and Narrative*, vol. 3, trans. Kathleen Blamey and David Pellauer. Chicago: University of Chicago Press.
Roberts, Christopher (2014). *The Contentious History of the International Bill of Human Rights*. New York and Cambridge: Cambridge University Press.
Rogers, Jack (1991). *Presbyterian Creeds: A Guide to the Book of Confessions*. Philadelphia: Westminster.
Rolin, Kristina (2008). 'Science as Collective Knowledge', *Cognitive Systems Research*, 9: 115–24.
Rolin, Kristina (2010). 'Group Justification in Science', *Episteme*, 7(3): 215–31.
Rovane, Carol (1998). *The Bounds of Agency*. Princeton: Princeton University Press.
Rudner, Richard (1953). 'The Scientist *qua* Scientist Makes Value Judgments', *Philosophy of Science*, 20(1): 1–6.
Sala-Molins, Louis (2006). *Dark Side of the Light: Slavery and the French Enlightenment*, trans. and with an introduction by John Conteh-Morgan. Minneapolis: University of Minnesota Press.
Salmela, Mikko (2012). 'Shared Emotions', *Philosophical Explorations*, 15: 33–46.
Sartre, J.-P. (1947). 'Conscience de soi et connaissance de soi'. Communication faite devant la Société Française de Philosophie, séance du 2 juin 1947.
Saul, Jennifer (2013). 'Implicit Bias, Stereotype Threat, and Women in Philosophy'. In *Women in Philosophy: What Needs to Change?*, ed. Katrina Hutchison and Fiona Jenkins. Oxford: Oxford University Press.

Schapiro, J. Salwyn (1963). *Condorcet and the Rise of Liberalism.* New York: Octagon Books.
Scherer, Klaus (1994). 'Emotion Serves to Decouple Stimulus and Response'. In Ekman and Davidson.
Schmid, Hans Bernhard (2009). *Plural Action: Essays on Philosophy and Social Science.* Dordrecht: Springer.
Schmid, Hans Bernhard (2014a). 'Plural Self-Awareness', *Phenomenology and the Cognitive Sciences*, 13(1): 7–24.
Schmid, Hans Bernhard (2014b). 'Expressing Group Attitudes: On First Person Plural Authority'. In *Erkenntnis*, 79 (DOI 10.1007/s10670-014-9635-8).
Schmid, Hans Bernhard, Daniel Sirtes, and Marcel Webe (2011). *Collective Epistemology.* Epistemische Studien: Schriften zur Erkenntnis- und Wissenschaftstheorie. Frankfurt a.M.: Ontos Verlag.
Schmitt, Frederick (1994). 'The Justification of Group Beliefs'. In *Socializing Epistemology: The Social Dimensions of Knowledge*, ed. Frederick F. Schmitt. Lanham, MD: Rowman & Littlefield: 257–87.
Searle, John R. (1990). 'Collective Intention and Action'. In *Intention in Communication*, ed. Sidney Pollack *et al.* Cambridge, MA: MIT Press: 404–15.
Setiya, Kieran (2011). 'Knowledge of Intention'. In *Essays on Anscombe's 'Intention'*, ed. Anton Ford, Jennifer Hornsby, and Frederick Stoutland. Cambridge, MA: Harvard University Press: 170–97.
Shapin, Steven (2008). *The Scientific Life: A Moral History of a Late Modern Vocation*, Chicago: University of Chicago Press.
Shaw, Malcolm Nathan (2003). *International Law.* Cambridge: Cambridge University Press.
Sher, George (2009). *Who Knew? Responsibility without Awareness.* Oxford: Oxford University Press.
Shoemaker, Sidney (1996). *The First-person Perspective and Other Essays.* Cambridge: Cambridge University Press.
Shue, Henry (1999). 'Global Environment and International Inequality', *International Affairs*, 75(3): 531–45.
Smith, Adam (1904). *An Inquiry into the Nature and Causes of the Wealth of Nations.* 5th. edn. London: Methuen & Co.
Smolin, Lee (2006). *The Trouble with Physics.* New York: Mariner Books.
Southwood, Nicholas (2011). 'The Moral/Conventional Distinction', *Mind*, 120(479): 761–802.
Spiekermann, Kai and Arne Weiss (forthcoming). 'Objective and Subjective Compliance: A Norm-Based Explanation of "Moral Wiggle Room"'. Working paper. Available at: <http://www.kaispiekermann.net/s/SpiekermannWeissMoralWiggle.pdf>.
Stanford, Kyle (2006). *Exceeding Our Grasp: Science, History, and the Problem of Unconceived Alternatives.* New York: Oxford University Press.
Stanley, Jason (2005). *Knowledge and Practical Interests.* Oxford: Oxford University Press.
Steup, Matthias, ed. (2001). *Knowledge, Truth, and Duty: Essays on Epistemic Justification, Responsibility, and Virtue.* New York: Oxford University Press.
Stilz, Anna (2011). 'Collective Responsibility and the State', *Journal of Political Philosophy*, 19: 190–208.

Sunstein, Cass (2006). *Infotopia: How Many Minds Produce Knowledge*. Oxford: Oxford University Press.
Sunstein, Cass and Reid Hastie (2015). *Wiser: Getting Beyond Groupthink to Make Groups Smarter*. Boston: Harvard Business Review Press.
Surowiecki, James (2005). *The Wisdom of Crowds: Why the Many Are Smarter Than the Few*. London: Abacus.
Swanton, Christine (1985). 'On the "Essential Contestedness" of Political Concepts', *Ethics*, 94: 811–27.
Taylor, Charles (1985). 'What Is Human Agency?' In *Human Agency and Language*. Cambridge: Cambridge University Press: 15–44.
Taylor, Charles (1993). 'Explanation and Practical Reason'. In *The Quality of Life*, ed. Martha Nussbaum and Amartya Sen. Oxford: Clarendon Press: 208–31.
Taylor, Charles (2004). *Modern Social Imaginaries*. Durham, NC: Duke University Press.
Thagard, Paul (1997). 'Collaborative Knowledge', *Noûs*, 31(2): 242–61.
Thagard, Paul (1999). *How Scientists Explain Disease*. Princeton: Princeton University Press.
Thompson, Janna (2002). *Taking Responsibility for the Past: Reparations and Historical Injustice*. Cambridge: Cambridge University Press.
Tilly, Charles (1993). 'Contentious Repertoirs in Great Britain, 1758–1834', *Social Science History*, 17(2): 253–80.
Tollefsen, Deborah (2002). 'Organizations as True Believers', *Journal of Social Philosophy*, 33: 395–410.
Tollefsen, Deborah (2007a). 'Collective Epistemic Agency and the Need for Collective Epistemology'. In *Facets of Sociality*, ed. Nikos Psarros and Katinka Schulte-Ostermann. Frankfurt a.M.: Ontos Verlag. 309–29.
Tollefsen, Deborah (2007b). 'Group Testimony', *Social Epistemology*, 21(3): 299–311.
Tracy, David (1996). *Blessed Rage for Order: The New Pluralism in Theology*. Chicago: University of Chicago Press.
Tuomela, Raimo (2006). 'Joint Intention, We-Mode and I-Mode', *Midwest Studies in Philosophy*, 30: 35–58.
Tuomela, Raimo (2007). *The Philosophy of Sociality: The Shared Point of View*. Oxford: Oxford University Press.
Tuomela, Raimo (2013). *Social Ontology: Collective Intentionality and Group Agents*. Oxford: Oxford University Press.
UK Government Archives, *Inquiries Acts 2005*: <http://www.nationalarchives.gov.uk/webarchive/public-inquiries-inquests.htm> <http://www.legislation.gov.uk/ukpga/2005/12/section/1>.
Unger, P. (1975). *Ignorance: A Case for Skepticism*. Oxford: Clarendon Press.
Vastey, Pompée Valentin, baron de (1818). *Political Remarks on Some French Works and Newspapers, Concerning Hayti*. London: The Pamphleteer.
Viefville des Essars, Jean Louis de (1790). 'On the Emancipation of the Negroes': <http://chnm.gmu.edu/revolution/d/340/> (accessed 29/8/2012).
Voyages Database (2009). 'Voyages: The Transatlantic Slave Trade Database'. At <http://www.slavevoyages.org/tast/database/search.faces>.
Wald, Abraham (1942). *On the Principles of Statistical Inference*. Notre Dame Mathematical Lectures, 1. Notre Dame, IN: University of Notre Dame.

Waldron, Jeremy (1999). *Law and Disagreement*. Oxford: Oxford University Press.
Walzer, Michael (1983). *Spheres of Justice: A Defense of Pluralism and Equality*. New York: Basic Books.
Weiner, M. (2005). 'Must We Know What We Say?', *Philosophical Review*, 114: 227–51.
Wendt, A. (2004). 'The State as Person in International Theory', *Review of International Studies*, 30: 289–316.
WHO (2015). *HIV Assays: Laboratory Performance and Other Operational Characteristics: Rapid Diagnostic Tests (Combined Detection of HIV-1/2 Antibodies and Discriminatory Detection of HIV-1 and HIV-2 Antibodies), Report 18*. Geneva: World Health Organization: <http://www.who.int/diagnostics_laboratory/publications/evaluations/en/> (accessed 25/3/2015).
Wilholt, Torsten (2009). 'Bias and Values in Scientific Research', *Studies in History and Philosophy of Science*, 40: 92–101.
Wilholt, Torsten (2013). 'Epistemic Trust in Science', *British Journal for the Philosophy of Science*, 64(2): 233–53.
Williams, Bernard (1982). 'Moral Luck'. In *Moral Luck: Philosophical Papers 1973–1980*. Cambridge: Cambridge University Press.
Williams, Bernard (1993). *Shame and Necessity*. Berkeley and London: University of California Press.
Williams, Rowan (1983). 'What Is Catholic Orthodoxy?' In *Essays Catholic and Radical*, ed. Rowan Williams and Kenneth Leech. London: Bowerdean: 11–25.
Williams, Rowan (1987). *Arius: Heresy and Tradition*. London: Darton, Longman & Todd.
Williamson, T. (1996). 'Knowing and Asserting', *Philosophical Review*, 105: 489–523.
Willis, David (1988). 'The Confession of 1967 in Historical Perspective', *Princeton Seminary Bulletin*, 9: 109–21.
Winsberg, Eric, Bryce Huebner, and Rebecca Kukla (2014). 'Accountability and Values in Radically Collaborative Research', *Studies in History and Philosophy of Science*, 46: 16–23.
Wittgenstein, Ludwig (1958). *Preliminary Studies for the 'Philosophical Investigations'*, generally known as *The Blue and Brown Books*. Oxford: Basil Blackwell.
Worrall, John (2002). 'What Evidence in Evidence-Based Medicine?', *Philosophy of Science*, 69 (Proceedings, S3): S316–30.
Wray, K. Brad (2001). 'Collective Belief and Acceptance', *Synthese*, 129(3): 319–33.
Wray, K. Brad (2002). 'The Epistemic Significance of Collaborative Research', *Philosophy of Science*, 69(1): 150–68.
Wray, K. Brad (2006). 'Scientific Authorship in the Age of Collaborative Research', *Studies in History and Philosophy of Science*, 37(3): 505–14.
Wray, K. Brad (2007). 'Who Has Scientific Knowledge?', *Social Epistemology*, 21(3): 337–47.
Wringe, Bill (2010). 'Global Obligations and the Agency Objection', *Ratio*, 23: 217–31.
Wolf, Susan (2013). 'The Moral of Moral Luck', *Philosophic Exchange*, Vol. 31(Issue 1), 1–16.
World Medical Association (2008). *Declaration of Helsinki*. At <http://www.wma.net/en/30publications/10policies/b3/>.
Zagzebski, Linda (1996). *Virtues of the Mind: An Inquiry into the Nature of Virtue and the Ethical Foundations of Knowledge*. Cambridge: Cambridge University Press.

Zagzebski, Linda (2001). 'Recovering Understanding'. In *Knowledge, Truth, and Duty*, ed. Matthias Steup. Oxford: Oxford University Press: 235–52.

Zagzebski, Linda (2004). *Divine Motivation Theory*. Cambridge: Cambridge University Press.

Zagzebski, Linda (2012). *Epistemic Authority: A Theory of Trust, Authority, and Autonomy in Belief*. Oxford: Oxford University Press.

Index

Aborigines, forced removal 164-5
Adams, R. M. 41 n.19
agent-regret 3-4, 44-50
altruistic behaviour 176-7, 187
Anderson, Elizabeth 186
Anglo Irish Bank 106
anonymization 47-8, 47-8 n.26, 49
Anscombe, G. E. M. 4, 51, 53-4, 57, 61
  approach to Collective Action 59-60, 65-6
Apostles' Creed 127
Australia:
  Aborigines and 164-5
  homosexual marriage 150, 162
  Pacific Solution policy 166-7

Bach, K. 14, 17
Bicchieri, Cristina 177, 179
Black Codes, American 88
blameworthiness 39 n.16, 39-41, 40 n.18, 42-3
Bratman, Michael 35
Brown, Joseph 86
Buchanan, Allen 182
Buechner, Frederick 122

Cappelen, Alexander W. 187
changing our minds 5, 111-29
  Christian theological ethics 125-8, 126 n.23
  Confession (1967), Presbyterian Church (US) 115, 116-17, 116 n.7, 124, 126-7
  current accounts of collective knowledge and belief revision 111-17
  from epistemology to collective moral epistemology 117 n.10, 117-20
  Grand Canyon example 117, 120-1, 123
  holistic knowledge-that 117, 123, 123 n.20
  Presbyterian Church (USA) example 114-17, 115 n.5, 115 n.6, 126-7, 128, 129
  propositional knowledge-that 117
  'Sprig of Hope' sermon 122, 122 n.19
  traditions, and collective belief revision 120 n.13, 120 n.15, 120-8
  understanding operator (u) 124
  virtue-epistemic account of knowledge 112-13, 113 n.3
  Westminster Confession (1646), Presbyterian 116, 124, 124 n.21, 126, 127

Chavez, Alex K. 177, 179
Christian theological ethics 125-8, 126 n.23
Church of Scotland 116
Churchman, C. West 228
climate change mitigation duty example 160-2
Cobb, Howell 86
Code Noir, France 79, 85
collective belief, and the string theory community 7, 191-217
  cognitive dissonance 200
  collective account 192-3, 193 n.3
  common knowledge 196, 196 n.11
  disinterest in other ideas, interpretation and risk 212-13
  Gilbert's account 195-201
  Grand Unified Theory 206
  identification with group, and boundaries 210-11, 211 n.31
  joint commitment 196-7, 198, 203
  Kuhnian paradigms 210, 215-16, 215-16 n.35
  Kuhnian paradigms and Gilbertian collective beliefs 201-4, 202 n.20
  M-theory 207-8
  quantum chromodynamics (QCD) 205
  rebukes and demands 197-8
  scientific communities 193-4, 194 n.5, 194 n.6, 199-201
  self-confidence and eliteness 213-14
  seven unusual features 214-16
  Smolin's sociological critique 209 n.29, 209-16
  Standard Model 205-6
  string theory, rise of 204-8, 208 n.26, 208 n.28
  summative account 192
  two-process views of scientific change 201
Condorcet, Nicolas 81-4, 119
Condorcet Jury Theorem (CJT) 140-1, 220
Confederate States, American 86
Confession (1967), Presbyterian Church (USA) 115, 116-17, 116 n.7, 124, 126-7
Cooperative Principle (Grice) 16-19, 22, 25, 27
Cuba 89
CV assessments 36-7

Dana, Jason 175-6, 181, 184, 185
Danton, Georges 86-7
D'Arm, Justin 102-3

Darwall, Stephen 143
De Grouchy, Sophie 84, 94
Declaration of the Rights of Man (1789), France 75
'deep proceduralist' conceptions 134
democracy, *see* epistemic circumstances of democracy 5-6, 133-49
Diana, Princess, death 95, 99, 100
dictator games 175-81, 176 Fig. 1, 182, 187
Douglas, Heather 228

Ellsworth, Phoebe 102 n.8
Emancipation Proclamation (1863), US 86
emotion, *see* group emotion, and group understanding
English Civil War 116
Enoch, David 48 n.27
epistemic circumstances of democracy 5-6, 133-49
 accountability thesis 146-7
 authority 134, 135
 authority dilemma 138, 140, 141, 149
 Condorcet Jury Theorem (CJT) 140-1
 'correctness theory' 137-8, 140-1
 deliberative democracy 135-6
 instrumentalism 133-4, 136, 137-41
 legitimacy 133, 134, 136, 137-41, 149
 legitimate authority 148-9, 149 n.12
 minimum-wage policy case 145-6
 peer disagreement 144-7, 145 n.10, 148
 peers 143-7, 145 n.9, 146 n.11
 practical and epistemic authority 136 n.3, 136-7
 procedural value of deliberation 141-7
 proceduralist approach 134-5
 third-personal authority 138, 139, 147, 148
equal-opportunities-awareness training 49-50
Estlund, David 137, 138-9
Evidence Based Medicine (EBM) 227

fault and no-fault responsibility 3-4, 33-50
 agent-regret 3-4, 44-50
 anonymization 47-8, 47-8 n.26, 49
 blameworthiness 39 n.16, 39-41, 40 n.18, 42-3
 de-biasing therapies 49-50
 epistemic culpability 34
 equal-opportunities-awareness training 49-50
 gender biases 42-3
 'implicit bias' that is implicit prejudice 36-43, 38 n.13
 institutional bodies 34-5
 lorry-driver case 45
 moral culpability 35
 moral luck 44, 45, 46

Oedipus case 45-6
'searchlight view' 41-2, 42 n.21
selfishness and blame 40-1
stereotyping 38-9
testimonial injustice 40, 40 n.17
Fetterolf, Elianna 45 n.23
Feynman, Richard 206
France:
 Code Noir 79, 85
 slavery 79-80, 80-5, 86-7
Francis, Robert 108 n.25, 108-9
Francis Report (2013), UK 108 n.25, 108-9
Fraser Inquiry (2003-2004), UK 106
Fricker, Miranda 183

Gallie, W. B. 125-6
Galton, Francis 119
gender biases 42-3
Gendler, Tamar 50
Georgi, Howard 206
German Association for Plant Biotechnology 222
Gettier, Edmund 112
Gibbard, Allan 118
Gilbert, Margaret 34-5, 111-12, 112 n.1, 113, 114, 115, 116, 117, 123, 128-9
 account of collective belief 5, 192-3, 193 n.3, 195-201
Glashow, Sheldon 206
Goldman, Alvin 227 n.10
Goodin, Robert 137
Grand Unified Theory, physics 206
Green, Michael 205, 206
Greene, Brian 208
Grégoire, Abbé 80
Grice, Paul 3, 16-19
 Cooperative Principle 16-19, 22, 25, 27
groundless group self-knowledge and plural self-blindness 4, 51-72
 'Anscombian approach to Collective Action' 59-60, 65-6
 breakdowns 52-3
 'by-product model' 69-70
 explained 52-8
 first-person:
  commitment 55-6, 60
  identity 54, 60
  perspective 54-5, 60
  plural commitment 63, 68-71
  plural identity 60-2, 66-7
  plural perspective 62, 67-8
 first-personal authority 57-8, 60, 63-5, 71-2
 'identification gap' 69-70
 illiism 64, 67
 joint action intentionality 58-65
 Moore's paradox 55-6, 63, 68
 plural pre-reflective self-awareness 65-72

pre-reflective self-awareness 65–6
spokespeople 64–5
group emotion, and group understanding 5, 95–110, 98 n.1
  'affective conformity' 99–100
  attentional persistence 101–2, 102 n.5, 102 n.6, 104, 104 n.14
  contagion 99, 100
  and affective conformity 106, 109
  individual and 96–101
    emotion and understanding 101–5, 101 n.4
  public inquiry cases 106 n.19, 106 n.20, 106–9, 107 n.21
  reflection and reappraisal 102–4
  tuition fees protest (2011), UK 98–9
Güth, Werner 180–1

Haisley, Emily C. 187
Haiti slave revolt 75, 87–8, 89, 90, 92
Harnisch, R. 17
Hastie, Reid 175
Heidelberg Catechism 127
Hempel, Carl 228
Henri Christophe, King of Haiti 90
HIV:
  placebo test 225
  tests, rapid 224–5, 225 n.9
Holroyd, Jules 34, 36, 37–9
homosexual marriage, Australia 150, 162
Huck, Steffen 180–1
Hume, David 135

Implicit Association Tests (IATs) 36–7
individual and state duties, *see* transfer of duties from individuals to states
intentional joint action theory 51

Jacobsen, Daniel 102
Jamaica 89, 90
Jefferson, Thomas 79, 80
Jesus of Nazareth 126

Kaku, Michio 208
Kelsen, Hans 133
King, Martin Luther 123
Konow, James 177–9, 178 n.2, 178 n.3, 187
Kuang, Jason Xi 175–7, 181, 184, 185
Kuhn, Thomas 7, 193, 193 n.4, 194–5, 201–4, 203 n.21, 216
Kuhnian paradigms 210, 215–16, 215–16 n.35
  and Gilbertian collective beliefs 201–4, 202 n.20
Kvanvig, Jonathan 117, 120 n.5, 120–1

Laming, Lord 107–8
Large Hadron Collider, CERN 219

Larmore, Charles 143
Laurence, Ben 59–60, 63
Lazarus, Richard 103–4
Leslie, Sara-Jane 38–9
Levenson Inquiry, UK 106
Lincoln, Abraham 91
List, Christian 35, 64–5, 69–70, 137
Locke, John 79, 80
lorry-driver case 45
Louis C. K. 55, 56
Louis XIV 79
Louverture, Toussaint 88

M-theory, physics 207–8
MacIntyre, Alasdair 129 n.26
Madva, Alex 50
marronage 84–5
Medina, José 183–4, 185, 186, 187
Mid Staffordshire NHS Foundation Trust, UK 108–9
Mill, John Stuart 94
Mills, Charles 184
Montevideo Convention on the Rights and Duties of States (1933) 158
Moore's paradox 55–6, 63, 68
moral luck 44, 45, 46
moral wriggle room 6–7, 173–88
  action-facts 173, 177, 181–2, 182 Tab. 1
  altruistic behaviour 176–7, 187
  'cognitive laziness' 185–6
  'epistemic laziness' 185
  epistemic neglect 183–4
  epistemology and racism 183–8
  four kinds 181–2, 182 Tab. 1
  'hiding behind small cakes' 180 n.6, 180–1, 182
  norm
    -fact acquisition 177–9
    -facts 173–4, 181–2, 182 Tab. 1
    perceptions 177–9
  obligations of inquiry 174–5
  strategic ignorance 175–7, 176 Fig. 1, 182 Tab.1
  strategic normative context shaping 174–5
  ultimatum game 179, 180
  'white ignorance' 184–5
Moran, Richard 57–8, 65
mutuality and assertion 2–3, 11–32
  assertoric practice and assessment 11–15, 13 n.5
  conditions 15–16, 19, 20, 22, 24, 25
  Conspirators group 29–31
  Cooperative Principle (Grice) 16–19, 22, 25, 27
  diminished epistemic hopes 12, 13, 14, 15, 20
  ethics and mutual belief 21–2
  high-grade belief 12, 12 n.3

mutuality and assertion (*cont.*)
  illocutionary force 16, 17
  norms 11, 13, 14, 16, 23
  objections considered 25-32
  'Quality' 16-17, 18, 25
  reasonable mutual belief norms (RMBNs) 27
  role for epistemic groups 19-25
  sincerity condition 13-14
  warranting authority 11, 18, 26-7

Napoleon Bonaparte 75
Nicene Creed 127

Ockenfels, Peter 180-1
O'Donovan, Oliver 121 n.16, 122 n.19
Oedipus case 45-6

Pacific Solution policy Australia 166-7
Paul, St. 124
Pettit, Philip 35, 64-5, 69-70, 156, 158, 159
Plato 133
Polverel, Étienne 87-8
Presbyterian Church (US) example 114-17, 115 n.5, 115 n.6, 126-7, 128, 129
Putnam, Hilary 121, 121 n.17

quantum chromodynamics (QCD) 205

racism 183-8
Randall, Lisa 208
randomized controlled trials (RCTs) 227
Rawls, John 125, 125 n.22, 135, 142 n.7, 142-3
Reid, Thomas 104
Republican Party, US 90-1
responsibility, *see* fault and no-fault responsibility
Rousseau, Jean-Jacques 79

Saint-Domingue 75, 80, 82-3, 86, 87-8
Sala-Molins, Louis 83
Sartre, Jean-Paul 65-6
Saul, Jennifer 34, 36
Scherer, Klaus 103
Scherk, Joël 205
Schmid, Hans Bernard 99-100
Schwarz, John 205, 206
Schweikard, David 156, 158, 159
scientific communities, and belief 193-4, 194 n.5, 194 n.6, 199-201
scientific research, and the social diffusion of trustworthiness 8, 218 n.1, 218-33, 232 n.14
  collaborative nature of 219-21, 220 n.5, 220 n.6
  distribution of inductive risks (DIR) 227-32, 228-9 n.13
  Evidence Based Medicine (EBM) 227

field specimen surveys 225-6
HIV:
  placebo test 225
  tests, rapid 224-5, 225 n.9
  investigation power 226-7, 227 n.10
  level of research communities 221-3, 222 n.8
  methodological standards 222-3
  methodological standards and distribution of inductive risks (DIR) 229-32
  randomized controlled trials (RCTs) 227
  reliability of methods 223-8
  significance tests 222, 227
Scots Confession 127
Scott v. Sandford 79
Scottish Parliament Building 106
Searle, John R. 66
Sher, George 41-2, 42 n.21
Shoemaker, Sydney 54-5, 55-6, 58, 67
Slavery Abolition Act (1833), Great Britain 75
slavery, abolition history 4-5, 75-94
  case for emancipation, testing theories of moral right 87-92
  Enlightenment proposals for gradual emancipation 79-84
  memory and moral epistemology 92-4
  slave participation in Enlightenment contention 84-7
Smith, Adam 84, 89-90
Smolin, Lee 7, 193-5, 206, 208
  sociological critique 209 n.29, 209-16
social epistemology of morality, and slavery 4-5, 75-94
  authoritarian moral enquiry 78-9
  case of emancipation, testing theories of moral right 87-92
  Enlightenment proposals for gradual slave emancipation 79-84
  'fundamental attribution error' 90-1
  history, memory, and moral epistemology 92-4
  sharecropping system 89
  slave participation in Enlightenment contention 84-7
Société des Amis des Noirs, France 81, 84
Sonthanax, Léger-Félicité 87-8
Spiekermann, Kai 177, 182
spokespeople 64-5
state/states:
  agents 158-9, 159 n.8
  'citizenship' 168
  'contracting out' 164-5
  exceptions, complications and clarifications 163 n.11, 163-5
  failure at multiple stages 165-7, 166 n.14
  membership 167-71, 169 n.16

nodes in the state story 159–62; see also transfer of duties from individuals to states
stereotyping 38–9
string theory:
　rise of 204–8, 208 n.26, 208 n.28
　see also collective belief, and the string theory community
Strominger, Andrew 211
Sunstein, Cass 175
Susskind, Leonard 208

Taylor, Charles 121 n.18
Taylor, Harriet 94
Thompson, William 94
Tollefsen, Deborah 112–13 n.3, 218
transfer of duties from individuals to states 6, 150–72
　capacity-relative duties 155
　citizen hermits 169, 171
　'citizenship' 168
　climate change mitigation duty example 160–2
　collective agent 156, 156 n.4, 156 n.5
　'collectivizing' 153–5, 154 n.3
　constituent roles and duties 156–8, 159, 172 n.18
　earthquake example 153–8
　homosexual marriage, Australia 150, 162
　nodes in the state story 159–62
　permanent resident alien status 169–70, 171
　political obligation 151
　the simple story 151, 152 Fig.1, 152–8, 153 n.1
　state:
　　agents 158–9, 159 n.8
　　exceptions, complications and clarifications 163 n.11, 163–5
　　failure at multiple stages 165–7, 166 n.14
　　membership 167–71, 169 n.16
　　states 'contracting out' 164–5
　　'we-thoughts' 157–8
Tuomela, Raimo 71, 112 n.1, 113, 120 n.13, 120 n.14

ultimatum game 179, 180
Universal Declaration of Human Rights (1948) 94
United States of America (US):
　slavery 80
　in the South 85–6

Vastey, Pompée Valentin, Baron de 90, 92
Velleman, J. D. 57
Victoria Climbié Inquiry (2001), UK 107–8, 108 n.25
Viefville des Essars, Baron de 80

Wald, Abraham 228
Waldron, Jeremy 135
Weber, Roberto A. 175–7, 181, 184, 185, 187
Weiss, Arne 177, 182
Westminster Confession (1646), Presbyterian 116, 124, 124 n.21, 126, 127
Wheeler, Anna Doyle 94
Williams, Bernard 33, 43–4, 45–6, 47
Williams, Rowan 126 n.25
Witten, Edward 207–8, 211
Wittgenstein, Ludwig 54
Wolf, Susan 48
women, equal rights 82, 82, 84; see also gender biases
Wray, K. Brad 192 n.2, 194 n.6, 194 n.7

Yoneya, T. 205

Zagzebski, Linda 117–18, 118 n.11

Printed and bound by CPI Group (UK) Ltd, Croydon, CR0 4YY